SpringerBriefs in Business Process Management

W0235191

Series Editor

Jan vom Brocke, Vaduz, Liechtenstein

More information about this series at http://www.springer.com/series/13170

Roope Jaakonmäki • Jan vom Brocke •
Stefan Dietze • Hendrik Drachsler •
Albrecht Fortenbacher • René Helbig •
Michael Kickmeier-Rust • Ivana Marenzi •
Angel Suarez • Haeseon Yun

Learning Analytics Cookbook

How to Support Learning Processes
Through Data Analytics and Visualization

 Springer

Roope Jaakonmäki
Department of Information Systems
University of Liechtenstein
Vaduz, Liechtenstein

Jan vom Brocke
Department of Information Systems
University of Liechtenstein
Vaduz, Liechtenstein

Stefan Dietze
L3S Research Center
University of Hannover
Hannover, Germany

Hendrik Drachsler (iD)
Welten Institute
Open University of the Netherlands
Heerlen, The Netherlands

Albrecht Fortenbacher
School of Computing, Communication
and Business
HTW Berlin
Berlin, Germany

René Helbig
School of Computing, Communication and Business
HTW Berlin
Berlin, Germany

Michael Kickmeier-Rust
KTI - Knowledge Technologies Institute
Graz University of Technology
Graz, Austria

Ivana Marenzi
L3S Research Center
Leibniz University of Hannover
Hannover, Germany

Angel Suarez
Welten Institute
Open University in the Netherlands
Heerlen, The Netherlands

Haeseon Yun
School of Computing, Communication and Business
HTW Berlin
Berlin, Germany

Disclaimer: "The book reflects only the authors' views. The Union is not liable for any use that may be made of the information contained herein."

ISSN 2197-9618   ISSN 2197-9626   (electronic)
SpringerBriefs in Business Process Management
ISBN 978-3-030-43376-5   ISBN 978-3-030-43377-2   (eBook)
https://doi.org/10.1007/978-3-030-43377-2

This Springer imprint is published by the registered company Springer Nature Switzerland AG.
The registered company address is: Gewerbestrasse 11, 6330 Cham, Switzerland

Preface

Rapidly advancing digitization is changing education dramatically. In response, educators need to adapt on a technical level—and, more important, on a pedagogical, conceptual level—how study material is distributed, how students interact in the classroom, how information is searched for and shared, and how learning scenarios are set up in general. The expansion of the web and the use of technology in education have provided not only new means to support learning and teaching but also the possibility of analyzing these learning processes through the digital traces students leave behind to improve teaching and learning.

Various research demonstrated that data have enormous potential for deep and novel insights into students' learning that can be used pedagogically, independent of a particular technology or medium. Gathering and analyzing digital traces from a learning environment can, for example, describe the students' abilities to learn during a certain teaching situation, identify issues that may hinder the learning experience, and predict which direction the students' learning path will take in the future. Visualizing the digital traces of teaching and training situations will make them more transparent so it is easier to see what could be improved. Accordingly, the high-level goals of learning analytics are to adapt teaching to the students' needs and to help identify the key indicators of students' performance in learning processes.

Data on each individual learner's strengths and weaknesses, learning paths, and gaps in competency can help educators support learners. The advantage educators have now that they did not have ten years ago is that there are now approaches that can support the difficult task of monitoring and evaluating learning activities using log data and visualizations of online activities. With an increased interest in such approaches, new and useful learning analytics tools and applications for various contexts and purposes have been developed.

However, solutions that are easy to implement are still sparse in the learning analytics community and among educators. It is often difficult for even the most enthusiastic teachers to build and implement learning analytics applications and derive meaning from the data on their own, especially in smaller organizations. This "cookbook" was created to showcase how easily learning analytics solutions could be implemented using any of a number of tools, just by following the steps in the "recipes."

Part I of the cookbook provides a general introduction to learning analytics, including its ethical aspects, but focuses particularly on the implications of the European data protection regulations. Part II of the book presents a set of learning analytics solutions that show how learning can be measured and how the data can be used in a pedagogically meaningful way. As in a cookbook, practical solutions are described and presented with step-by-step instructions.

The book is a starting point for collecting a large number of learning analytics recipes on a web platform [http://www.learning-analytics-toolbox.org/]. A collection of such recipes in the same database is the most convenient way to make a wide range of learning analytics applications globally available and easy to use. With a rational way of categorizing the recipes and providing filtering options, anyone could search and try out learning analytics based on a specific objective, a given amount of resources, and a certain context and complexity level.

We hope that this cookbook leverages initiatives that have been undertaken to bring the most recent developments and research results in the context of learning analytics into practice so learners and educators on all levels can benefit from modern data analytics and visualization technologies.

Vaduz, Liechtenstein	Roope Jaakonmäki
Vaduz, Liechtenstein	Jan vom Brocke
Hannover, Germany	Stefan Dietze
Heerlen, The Netherlands	Hendrik Drachsler
Berlin, Germany	Albrecht Fortenbacher
Berlin, Germany	René Helbig
Graz, Austria	Michael Kickmeier-Rust
Hannover, Germany	Ivana Marenzi
Heerlen, The Netherlands	Angel Suarez
Berlin, Germany	Haeseon Yun

Acknowledgments

This book presents the results of a European Project funded under the schema of ERASMUS+ (2015-1-LI01-KA203-000041). We are grateful for ERASMUS+'s support, which has been vital in creating this consortium and producing results that can now be applied and further developed by colleagues around the world. While Roope Jaakonmäki from the University of Liechtenstein served as the project coordinator and ensured that we succeeded in creating this cookbook, the results are a joint effort to which all coauthors contributed. Therefore, Jaakonmäki is the first author, and all other authors are listed in alphabetic order.

We also thank several people who helped in making this book possible. Special thanks go to Peter Sommerauer, who initiated the project, and to Clarissa Frommelt from National Agency LI01—Agentur für Internationale Bildungsangelegenheiten (AIBA)—for her valuable support during the project. We are also deeply indebted to Guido Forster, Michael Gau, Shaho Alaee, Magdalena Eggarter, Alin Secareanu, and Dr. Sonia Lippe-Dada, all of whom contributed to various phases of the project. We thank Dr. Miroslav Shaltev, who installed the first Wordpress instance on the Server in Hannover, and Rubayet Hasan, who initiated the first version of the website. Our thanks also go to Dr. Besnik Fetahu and Dr. Asmelash Teka Hadgu, who contributed to the identification and description of some of the learning analytics datasets. Finally, we thank the teacher and the students of the University of Salento, Lecce, Italy, for providing a real teaching scenario.

Contents

Glossary 27

Part I
Understanding Learning Analytics

Chapter 1
Introduction to the Learning Analytics Cookbook

Abstract Many stakeholders, including educators, researchers, university administrators, and others might be interested in implementing learning analytics solutions, but they have only limited experience with it, lack the technical expertise or funding, or simply don't know where to begin. This cookbook brings learning analytics a step closer to educators, smaller organizations, and educational institutions by presenting approaches that do not require complex or expensive infrastructures. The first chapter gives a brief overview of what the Cookbook and learning analytics recipes are all about. In addition, the chapter gives introduction to the learning analytics Toolbox concept. The concept is an initiative to organize all learning analytics materials and guidelines (e.g., objectives, methods, tools, datasets, and recipes) in a compact and structured way to appeal to a variety of audiences interested in learning analytics. This, hopefully, further motivates the learning analytics community to develop one centralized web portal for learning analytics.

Keywords Cookbook · Recipes · Toolbox

1.1 Cookbook and Recipes

The goal of this cookbook is first to explain the concept of learning analytics and then to provide a collection of small, ready-to-use, hands-on experiences for educational institutions that are interested in what learning analytics might be able to do for their practices. So how could we summarize what learning analytics actually is? Duval and Verbert (2012) defined learning analytics as the process of collecting traces of information from learners and analyzing them to improve learning. These traces are harvested, processed, modeled, and offered in the form of visualization dashboards that help learners and teachers make informed decisions about current and future learning performance. The complete cycle of harvesting to visualization and prediction to the implementation of learning analytics techniques involves complex endeavors that are today accessible only to schools and institutions with entire departments that are dedicated to learning analytics.

© The Author(s) 2020   3
R. Jaakonmäki et al., *Learning Analytics Cookbook*, SpringerBriefs in Business
Process Management, https://doi.org/10.1007/978-3-030-43377-2_1

Other books (e.g., Lang et al. 2017; Larusson and White 2014) offer introductions to the topic and present the current state of the learning analytics field, with examples and use cases of how to enhance digital learning, teaching, and training at various levels. However, applying these approaches yourself at your institute may be difficult for any number of reasons, but especially because of the complexity of the topic. This book bridges this complexity gap by providing ready-to-use applications of learning analytics in a recipe format that makes it easier to follow the steps to implement learning analytics.

We start with a discussion of the learning analytics "kitchen", which represents the most important themes that researchers, practitioners, and legal entities deal with, including the ethical issues involved, data privacy, and data protection. Users of learning analytics must be aware of legal and ethical boundaries to avoid abusing sensitive data and to ensure that data collection and handling processes have the required transparency.

In the second part of the book, we offer hands-on experiences with learning analytics. These are "served" in the form of recipes, with information about the required ingredients, preparation, cooking, and serving. These recipes not only serve as examples of learning analytics scenarios but also offer the opportunity to start experimenting with learners' digital traces. The recipes offer the required information and tools and guide the reader through the steps to achieving a specific learning analytics goal, such as how to visualize the students' activity with respect to a given learning situation in a digital environment.

Each recipe consists of five different phases:

1. Appetizer—Putting the chef into the context with regard to the specific learning analytics opportunity, the students, and the environment.
2. Ingredients—The resources needed to complete the recipe, referring to the data and technologies required.
3. Preparation—How to configure and set up the environment for cooking the recipe.
4. Cooking—Guides the reader through the recipe's steps to accomplish the goal of gaining insights into the specific learning situation.
5. Serving—A concrete example of the final output and the value added by the recipe.

1.2   The Learning Analytics Toolbox

The Learning Analytics Toolbox works as a supplement to the cookbook. While the cookbook offers an introduction to learning analytics and a set of easy-to-implement solutions for learning analytics needs, the Toolbox is contains more detailed information about common objectives and methods and shares more insights into data protection and legal regulations in Europe. Having a sound understanding of these topics makes it easier to plan, develop, and apply learning analytics activities in practice.

The Toolbox also describes various publicly available datasets for learning analytics and points out the sources for downloading them. Currently, publicly available anonymized datasets online that describe students' behavior are scarce, and the datasets' natures vary widely. For example, some can be downloaded for the purpose of getting a better understanding of what raw or only lightly preprocessed data looks like, while others can be for experimenting with converting the data into valuable information or for various research purposes. The "Open University Learning Analytics dataset" is a good example to take a look at to get an idea of what kind of data could be used to analyze students' activities and performance, as it describes students' course performance and their interactions with virtual learning environments (Kuzilek et al. 2017).

Various tools on the market can help with the analyses and visualizations of data collected from learners. Many of these tools have multiple functionalities, not all of which might be needed in the end. Some of these software products also require paying license fees, so they might not even be an option for many schools and universities, especially the smaller ones that have fewer resources. However, applying learning analytics is possible without buying these complete solutions, so the Toolbox points out some tools that could be good options to try out even when resources are limited. These materials and guidelines are available through an online portal (http://www.learning-analytics-toolbox.org/), which is designed to appeal to a variety of audiences, including teachers, researchers, and learners (Fig. 1.1).

Fig. 1.1 Learning analytics toolbox home screen

References

Duval, E., & Verbert, K. (2012). Learning analytics. *ELEED: E-Learning and Education, 8*(1).

Kuzilek, J., Hlosta, M., & Zdrahal, Z. (2017). Open university learning analytics dataset. *Scientific Data, 4,* 170171.

Lang, C., Siemens, G., Wise, A., & Gasevic, D. (Eds.). (2017). *Handbook of learning analytics.* Beaumont, AB: Society for Learning Analytics and Research (SOLAR).

Larusson, J. A., & White, B. (Eds.). (2014). *Learning analytics: From research to practice.* New York: Springer.

Chapter 2
Learning Analytics Kitchen

Abstract Among trending topics that can be investigated in the field of educational technology at present—Massive Open Online Courses (MOOCs), Open Educational Resources (OERs), wearables, and so on—there is a high demand for using educational data to improve the whole learning and teaching cycle, from collecting and estimating students' prior knowledge of a subject to the learning process and its assessment. Therefore, educational data science cuts through almost all educational technology disciplines. Before we can start applying learning analytics, it is good to have more holistic understanding of learning analytics and its surrounding environment. This chapter describes the field of learning analytics and recent research to give a better overview of the concept.

Keywords Framework · Dimensions · Scope

2.1 Defining the Learning Analytics Niche

Using data to inform decision-making in education and training is not new, but the scope and scale of its potential impact on teaching and learning has increased by orders of magnitude over the last few years. We are now at a stage at which data can be automatically harvested at high levels of granularity and variety. Analysis of these data has the potential to provide the evidence-based insights into learners' abilities and patterns of behavior that can guide curriculum design and delivery to improve outcomes for all learners and change assessments from mainly quantitative and summative to more qualitative and formative, thereby contributing to national and European economic and social well-being.

Data science extracts knowledge and insights from data, and data science in the educational context is termed *learning analytics*. Learning analytics is an umbrella term for data-science-related research questions from overlapping fields like the educational, computer, and data science fields. The umbrella term has many facets and can be described from many levels and angles.

Learning analytics can be applied to the single-course level, a collection of courses, or a whole curriculum. Buckingham Shum (2012) introduced the notion

R. Jaakonmäki et al., *Learning Analytics Cookbook*, SpringerBriefs in Business Process Management, https://doi.org/10.1007/978-3-030-43377-2_2

of micro-, meso- and macro-levels (Fig. 2.1) to distinguish the roles that learning analytics can play on different abstraction levels. The micro-level mainly addresses the needs of teachers and students in the context of a single course; the meso-level addresses a collection of courses and provides information for course managers; and the macro-level takes a bird's-eye view of a whole curriculum by monitoring learning behavior across courses and even across disciplines. Depending on the level at which learning analytics takes place, different objectives are set, so the information sought also differs.

Although learning analytics has been around for several years, and various start-up companies have provided learning analytic tools (with the support of plenty of venture capital), most of the strategies are still in their initial phase of gaining awareness (Fig. 2.2) because, despite the current enthusiasm about learning analytics, there are substantial questions for research and organizational development that have stagnated its implementation (Siemens et al. 2013). In some prominent cases, concerns by governments, stakeholders, and civil rights groups about privacy and

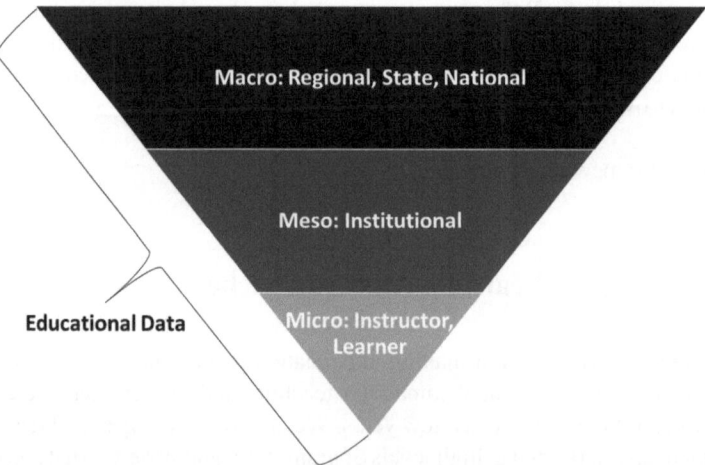

Fig. 2.1 Levels of learning analytics. Adapted from "Learning Analytics," by S. Buckingham Shum (2012), IITE Policy Brief, p. 3. Copyright 2002 by UNESCO

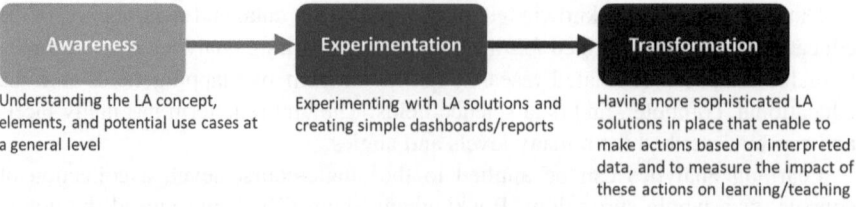

Fig. 2.2 Stages of learning analytics adoption

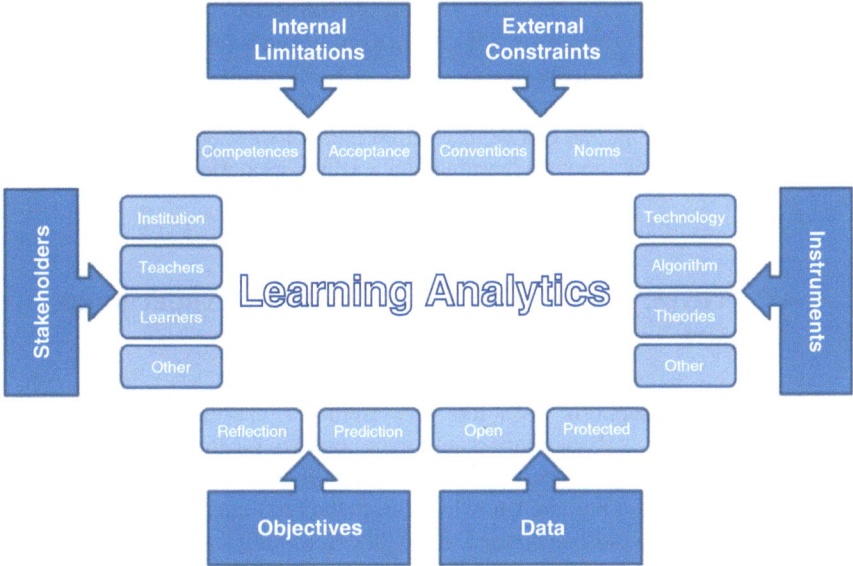

Fig. 2.3   Six dimensions of learning analytics (Greller and Drachsler 2012)

ethics applied to the handling of personal data have even reversed implementations (Singer 2014; Tsai et al. 2018).

Greller and Drachsler (2012) provided a comprehensive introduction to the domains affected by learning analytics (Fig. 2.3). They clustered learning analytics into six dimensions: stakeholders, objectives, data, instruments, external constraints, and internal limitations. These dimensions are based on the assumption that developers of analytic processes will implement what is technically possible and legally allowed but also consider the outcomes for the educational stakeholders and, even more important, the consequences for the people whose data is being used.

2.2   Stakeholders: Contributors and Beneficiaries of Learning Analytics

The stakeholder dimension includes *data clients* as well as *data subjects*. Data clients are the beneficiaries of the learning analytics process who are meant to act upon the outcome (e.g., students and teachers). The data subjects are those who supply data, normally through their browsing and interaction behavior. These roles can change depending on the objective of the analytics (e.g., whether it is on the micro-, meso-, or macro-level). Moreover, educational technology requires a sound analysis of stakeholders' needs to be successful in the long run (Greller and Drachsler 2012). This requirement applies especially to learning analytics solutions, as data sources are often highly diverse, and the European countries' educational systems are highly heterogeneous, sometimes even within a single country, because of the countries'—and sometimes their states' (e.g., Germany)—autonomy. Therefore, a one-size-fits-all approach is far inferior to a sophisticated methodology that identifies key indicators that can be transferred into software engineering diagrams and early prototypes to innovate learning and teaching (Scheffel et al. 2014).

2.3   Objectives: Goals That Learning Analytics Applications Support

Learning analytics supports tasks like measurements, analyses, and predictions (based on digital data) about learners and their contexts for the purpose of understanding and optimizing the learning situation and the learning environment (statement on Learning Analytics and Knowledge from the First International Conference in 2011). In other words, supported by digital technologies, learning analytics focuses on improving teaching and learning activities in learning environments and making learning and teaching situations more transparent to both teachers and students.

That said, there is a wide range of incentives for educators to experiment with learning analytics. Figure 2.4 illustrates the frequency of learning analytics objectives addressed in the learning analytics literature based on how many times a set of predefined objectives (derived from Baker and Siemens 2015; Chatti et al. 2012;

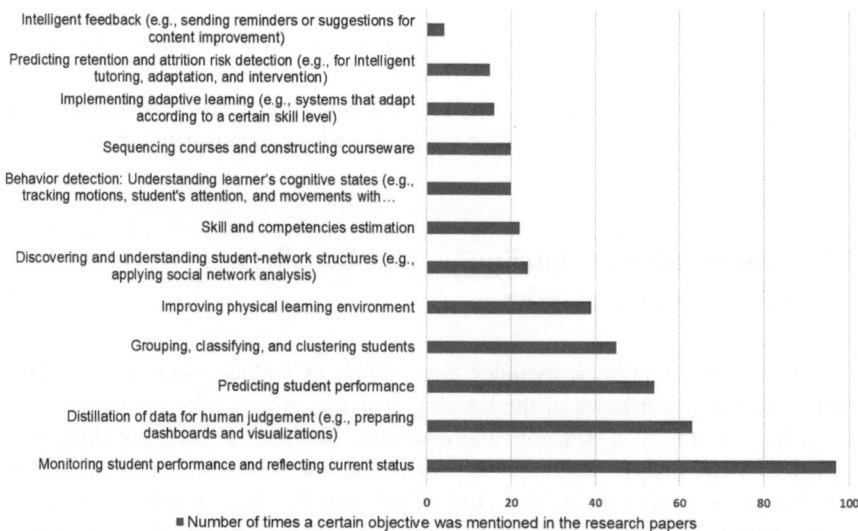

Fig. 2.4 Frequency of objectives addressed in the learning analytics literature (401 research papers)

Greller and Drachsler 2012; Nistor et al. 2015; Papamitsiou et al. 2014; Sin and Muthu 2015) occurred in 362 papers published in *Learning Analytics & Knowledge Conference* and 39 papers in *Journal of Educational Measurement* between 2010 and 2016. More detailed descriptions of each objective are available in the Learning Analytics Toolbox at http://www.learning-analytics-toolbox.org/home/objectives/.

The most common objective is to monitor the student's performance. This refers to tracking how active students are and how they perform during a course or their studies. The second most common objective in the core of learning analytics literature is to distill and visualize various patterns in students' behavior that could improve learning and teaching activities via better understanding of the learning situation. It is clear that these two objectives (Fig. 2.4) are broader than the others. When learning analytics are applied in practice, the first objective is usually something other than just making visualizations (e.g., classifying students into certain groups, predicting students' performance). For example, the goal of applying learning analytics could be to estimate students' competencies and then to create a dashboard that presents insights from the data in a convenient and comprehensible way. However, one could also just estimate students' competencies without creating sophisticated visualizations or dashboards.

Many stakeholders are probably interested in implementing learning analytics solutions, but it is not easy to know where to start. An overview of these objectives indicates the types of objectives learning analytics could address. Understanding the potential areas of implementation makes identifying the tools and data that are available in a certain teaching or learning environment easier.

After an objective for applying learning analytics is clear, it still might not be easy to choose the right method to meet it. Learning analytics applies known methods and models that have been used in other types of analytics to address issues that affect students' learning processes and the organizational learning system. The list of learning analytics methods is long (and would be longer if all of the variations of such methods were taken into account). For a high-level view of what kind of methods are out there and what they are good for, we present a list of methods and present them in the Learning Analytics Toolbox (http://www.learning-analytics-toolbox.org/methods/).

2.4 Data: Educational Datasets and the Environments in Which They Occur

Typical scenarios for learning analytics include e-learning courses and perhaps popular MOOCs, where a learner produces data with every mouse click. However, students usually do their exercises and homework with paper and pencil, which matches the social, interactive, and personal process of learning, an analogue process. Because of the scarcity of digital traces from students, we rarely have access to educational "big data" and sometimes do not even find "little data," as sometimes the only digital data is a final grade. Teachers also tend to build their appraisals of students in an intuitive and experience-based way instead of an objective data-based, evidence-based way. More data would be the key to creating more holistic profiles of the learners, which requires collecting information about learners from various sources and storing, interpreting, and aggregating it.

For example, learning analytics uses datasets from institutions that are protected from external access and use. However, an increasing number of open and linked data sources from governments and organizations like the Organisation of Economic Cooperation and Development (OECD) can be used to investigate target groups for certain courses or programs (d'Aquin et al. 2014). Among the providers and users of these closed and open datasets is a movement toward more standardized metadata for learning analytics (i.e., new specifications for learning technology and frameworks like xAPI[1] and IMS Caliper[2]). The use of such metadata standards allows data to be combined and results gained in various scientific disciplines and educational scenarios to be compared (Berg et al. 2016). A comprehensive uptake of such data could lead to a paradigm shift in educational science, a field that is currently more accustomed to small-scale experimental studies than to big-data-driven ones like those performed at Google and Facebook (Kramer et al. 2014). Apart from strengthening the research side with standardized educational metadata and being able to convert this data into one unified format, such standards could also create a market space for educational tools and open educational resources and their analytics.

2.5   Instruments: Technologies, Algorithms, and Theories That Carry Learning Analytics

Several technologies can be applied to the development of educational services and applications that support educational stakeholders' objectives (Drachsler et al. 2015). Learning analytics takes advantage of machine learning (where computers can "learn" from patterns), analysis of social networks, and classical statistical analysis and visualization techniques (Fazeli et al. 2014; Scheffel et al. 2017a). Through these technologies, learning analytics can contribute tailored information to stakeholders and report on demand. For instance, learning analytics can be applied to develop a system that identifies students who are in danger of dropping out.

---

[1]http://tincanapi.com/

[2]http://imsglobal.org/caliper/index.html

2.6 Internal Limitations: User Requirements to Exploit the Benefits of Learning Analytics

Learning analytics does not end with the presentation of algorithmically attained results, as those results require interpretation, which entails high-level competencies, such as interpretative and critical evaluation skills. Since these skills are not currently standard for stakeholders in the educational field, another challenge for the adoption and rollout of learning analytics is to increase the competencies of the educational stakeholders, mainly teachers and students. Educational technology should also take a leading role in teacher and learner training to increase those competencies. How well learning analytics address the needs of the students and teachers can be measured with the Evaluation Framework for Learning Analytics (EFLA), an empirical, data–based evaluation instrument (Scheffel et al. 2014, 2017a, b) that is freely available at the LACE project website.[3]

2.7 External Constraints: Restrictions and Limitations to Anticipated Benefits

The large-scale production, collection, aggregation, and processing of information from educational programs have led to ethical and privacy concerns regarding the potential for harm to individuals and society. How apt such concerns really are has become evident in prominent examples like the shutting down of inBloom in the United States because of privacy concerns related to learning analytics and big data in education (Singer 2014). Few papers have been published that relate to the ethics of learning analytics and privacy and potential solutions, but first policies and guidelines regarding privacy, legal protection rights, and ethical implications have been announced, such as the policy published by the Open University UK (2014). Drachsler and Greller (2016) investigated the most common fears and propositions concerning privacy and ethics and concluded with an eight-point checklist named DELICATE that researchers, policy-makers, and institutional managers can apply to facilitate a trustworthy implementation of learning analytics (See also Sect. 3.3 on ethical frameworks in learning analytics).

References

Baker, R., & Siemens, G. (2015). Educational data mining and learning analytics. In R. K. Sawyer (Ed.), *The Cambridge handbook of the learning sciences* (2nd ed., pp. 253–274). Cambridge: Cambridge University Press.

---

[3]http://www.laceproject.eu/evaluation-framework-for-la/

Berg, A., Scheffel, M., Drachsler, H., Ternier, S., & Specht, M. (2016). Dutch cooking with xAPI recipes: The good, the bad, and the consistent. In *2016 IEEE 16th international conference on advanced learning technologies (ICALT)* (pp. 234–236). Austin, TX: IEEE.

Buckingham Shum, S. (2012). *Learning analytics. UNESCO IITE policy brief*. Retrieved from http://iite.unesco.org/files/policy_briefs/pdf/en/learning_analytics.pdf

Chatti, M. A., Dyckhoff, A. L., Schroeder, U., & Thüs, H. (2012). A reference model for learning analytics. *International Journal of Technology Enhanced Learning, 4*(5-6), 318–331.

d'Aquin, M., Dietze, S., Herder, E., Drachsler, H., & Taibi, D. (2014). Using linked data in learning analytics. *eLearning papers*, 36. Retrieved from http://www.openeducationeuropa.eu/en/down load/file/33993

Drachsler, H., & Greller, W. (2016). Privacy and analytics: It's a DELICATE issue a checklist for trusted learning analytics. In *Proceedings of the sixth international conference on learning analytics & knowledge (LAK '16)* (pp. 89–98). New York: ACM.

Drachsler, H., Verbert, K., Santos, O. C., & Manouselis, N. (2015). Panorama of recommender systems to support learning. In F. Rici, L. Rokach, & B. Shapira (Eds.), *2nd handbook on recommender systems* (pp. 421–451). New York: Springer.

Fazeli, S., Loni, B., Drachsler, H., & Sloep, P. (2014). Which recommender system can best fit social learning platforms? In *9th European conference on technology enhanced learning (EC-TEL 2014)* (pp. 84–97). Graz, Austria: Springer.

Greller, W., & Drachsler, H. (2012). Translating learning into numbers: A generic framework for learning analytics. *Educational Technology & Society, 15*, 42–57.

Kramer, A. D., Guillory, J. E., & Hancock, J. T. (2014). Experimental evidence of massive-scale emotional contagion through social networks. *Proceedings of the National Academy of Sciences, 111*(24), 8788–8790.

Nistor, N., Derntl, M., & Klamma, R. (2015). Learning analytics: Trends and issues of the empirical research of the years 2011-2014. In *Proceedings of the 10th European conference on technology enhanced learning* (pp. 453–459). Toledo, Spain: Springer.

Open University UK. (2014). *Policy on ethical use of student data for learning analytics*. Retrieved from http://www.open.ac.uk/students/charter/sites/www.open.ac.uk.students.charter/files/files/ecms/web-content/ethical-student-data-faq.pdf

Papamitsiou, Z. K., Terzis, V., & Economides, A. A. (2014). Temporal learning analytics for computer based testing. In *Proceedings of the fourth international conference on learning analytics and knowledge* (pp. 31–35). Indianapolis: ACM.

Scheffel, M., Drachsler, H., Stoyanov, S., & Specht, M. (2014). Quality indicators for learning analytics. *Journal of Educational Technology & Society, 17*(4), 117.

Scheffel, M., Drachsler, H., de Kraker, J., Kreijns, K., Slootmaker, A., & Specht, M. (2017a). Widget, widget on the wall, am I performing well at all? *IEEE Transactions on Learning Technologies, 10*(1), 42–52.

Scheffel, M., Drachsler, H., Toisoul, C., Ternier, S., & Specht, M. (2017b). The proof of the pudding: Examining validity and reliability of the evaluation framework for learning analytics. In *European conference on technology enhanced learning* (pp. 194–208). Berlin, Germany: Springer.

Siemens, G., Dawson, S., & Lynch, G. (2013). Improving the quality and productivity of the higher education sector – policy and strategy for systems-level deployment of learning analytics. Sydney, Australia. Retrieved from http://solaresearch.org/Policy_Strategy_Analytics.pdf

Sin, K., & Muthu, L. (2015). Application of big data in education data mining and learning analytics: A literature review. *ICTACT Journal on Soft Computing, 5*(4), 1035–1049.

Singer, N. (2014, April 21). InBloom student data repository to close. *New York Times*. Retrieved from https://bits.blogs.nytimes.com/2014/04/21/inbloom-student-data-repository-to-close/

Tsai, Y. S., Moreno-Marcos, P. M., Tammets, K., Kollom, K., & Gašević, D. (2018). SHEILA policy framework: Informing institutional strategies and policy processes of learning analytics. In *Proceedings of the 8th international conference on learning analytics and knowledge* (pp. 320–329). Sydney, Australia: ACM.

Chapter 3
Responsible Cooking with Learning Analytics

Abstract Misuse of students' data may cause significant harm. Educational institutions have always analyzed their students' data to some extent, but earlier that data was stored locally and in analog form. The digital revolution has led to more data being generated, stored, and linked, and more conclusions drawn from it. These new possibilities demand a new code of conduct for generating, processing, using, and archiving student data. It is crucial that ethical aspects are carefully considered before collecting and using such data in learning analytics scenarios. This chapter gives an overview of legal regulations, illustrates the critical aspects of working with digital student data, and provides practical frameworks for ethics and data protection. This chapter includes parts of legislations and guidelines that were presented in a published article: *Steiner, C. M., Kickmeier-Rust, M. D., & Albert, D. (2016). LEA in private: a privacy and data protection framework for a learning analytics toolbox. Journal of Learning Analytics, 3(1), 66-90.* This chapter updates and extends the original article.

Keywords Data protection · Ethics · Guidelines

3.1 The Responsible Learning Analytics Kitchen

Today's learners have access to a broad range of digital devices, learning tools, applications, and resources. Learning experiences occur in virtual and simulated environments (e.g., serious games), and students connect to each other in social networks, so the lines between the use of technology in educational settings, whether in schools or in the field of workplace learning, and private settings are increasingly blurred. All of these interactions and links could be captured, and multifaceted learning processes can be analyzed using big-data analytics techniques (Pardo and Siemens 2014). However, the increasing digitization of education, the increasing capacity and adoption of learning analytics, and the increasing linkage of various data sources have created ethical and privacy issues. An example is the advancement in sensor technologies, wearables, and near-field communication technologies that allow a person's activities, locations, and contextual data to be tracked without the person's even being aware of it. Data collection and use under such circumstances is, of course, ethically and legally questionable (Greller and Drachsler 2012).

Learning analytics is the mechanism that brings such perils of digitization into education, classrooms, and lecture halls. Often, educators are either not fully aware of these issues when they apply various technologies or they are over-anxious and hesitant to use them. Ethical and data protection issues in learning analytics include the collection of data, informed consent, privacy, de-identification of data, transparency, data security, interpretation of data, data classification and management, and potential harm to the data subject (cf. Sclater 2014a; Slade and Prinsloo 2013). A number of researchers have worked on establishing an agreed set of guidelines with respect to the features of learning analytics, such as the ownership of data and analytic models and rights and responsibilities (Berg 2013; Ferguson 2012).

The more interlinked, multimodal, and multifaceted learning analytics becomes, and the more frequent and widespread its use, the more important are issues related to privacy and data protection, particularly since the introduction of the 2018 EU General Data Protection Regulation (GDPR). These strict regulations and rules establish a specific code of conduct when applying learning analytics in educational settings. The following nine categorizations and dimensions can guide users through the process of integrating solid data protection and privacy measures into their learning analytics applications (Sclater 2014a; Slade and Prinsloo 2013; Willis 2014).

Privacy Privacy refers to "the quality or state of being apart from company or observation" (Merriam Webster Dictionary 2019). Aggregating and analyzing personal data may violate this principle. An example is that learning analytics may allow teachers to view all of an individual's working steps, rather than just the assigned outcome.

Transparency Transparency refers to informing learners about all of the features, processes, and mechanisms of the application of learning analytics and to making the often hidden analytics processes and means of automatic grading understandable to the learners.

Informed Consent Informed consent refers to the process of ensuring that individual learners understand the features, processes, and mechanisms of the application of learning analytics and the potential consequences and to seeking the consent of individual learners (or their legal representatives) to apply learning analytics.

De-identification of Data Data de-identification refers to separating performance data from data that can identify a particular individual.

Location and Interpretation of Data Educational activities that are tracked by learning analytics are usually spread over multiple tools and locations. Learning analytics brings these data sources together for a more complete picture of learning. Questions about the implications of using multiple and perhaps noninstitutionalized sources arise concerning whether the data is representative of a particular student.

Data Management, Classification, and Storage Data management, classification, and storage refer to questions about data management, access rights, the measures and level of data protection needed, and how long the data can be stored.

Data Ownership The question regarding who is the owner of digitally produced data, its analyses, and outputs in the education context is linked to the outsourcing and transfer of data to third parties, and related regulations and responsibilities.

Possibility of Error Analytics results are always based on the data available, so the outputs and predictions obtained may be imperfect or incorrect. Therefore, questions about the ramifications of making an error and the implications of ineffective or misdirected interventions have arisen.

Obligation to Act Learning analytics may provide new knowledge about and insights into learning. Therefore, the question arises concerning whether new knowledge entails the responsibility to act on it, bearing in mind that the results of analyses may be influenced by improper algorithms and concepts or incomplete or erroneous data.

3.2   Big Data in the Responsible Learning Analytics Kitchen

Privacy and ethics are issues in all areas of data science, such as big data analytics (e.g., Richards and King 2014). As Tene and Polonetsky (2013, p. 251) emphasized, "big data poses big privacy risks." Data has become a resource of significant economic and social value, and the exponentially growing amount of data from a multitude of devices and sensors, digital networks, and social media that is generated, shared, transmitted, and accessed, together with newly available technologies and analytics, opens up new and unanticipated uses for information. The disclosure and use of personal data are increasingly associated with fear, uncertainty, and doubt (Anderson and Gardiner 2014). According to UN privacy chief Joseph Cannataci, digital surveillance is "worse than Orwell" (as cited in Alexander 2015).

Users are concerned about their privacy in settings where data is collected untrammeled and that large amounts of their personal information may be tracked and made accessible to other users for unknown purposes. A counter-effect in society is that, because of the ubiquity of data-collecting technologies in the people's daily lives and routines, people have become increasingly desensitized to it (Debatin et al. 2009) and are willing to share many personal details via these networks. Attitudes toward privacy often differ from privacy-related behaviors (Stutzman and Kramer-Duffield 2010), leading to the "privacy paradox" (Barnes 2006). Debatin et al. (2009) supported this view by comparing users' confident self-reports about their understanding of the dangers related to privacy settings and their unconcerned behavior in the form of retaining default settings instead of updating them to reflect their needs and preferences.

Therefore, people may not act according to the privacy preferences they claim, and they are usually unconcerned about data protection and privacy until their data are breached (Spiekerman and Cranor 2009). Users' concerns about privacy also depend on the type of data that are collected, the context within which they are collected, and their perceptions of the dangers in disclosing them (Pardo and Siemens 2014). The problem with this paradox effect in terms of learning analytics is that designers, developers, educators, and learners tend either to underestimate or overestimate the effects of and the importance to data subjects of privacy and data protection.

Tene and Polonetsky (2013) investigated the fundamental principles of privacy codes and legislation and argued that the principles of data minimization and individual control and context should to be somewhat relaxed in a big-data context, where individual data points are not connected to individual data subjects, because of the potential benefits to society (e.g., public health, environmental protection) and, along with society, themselves as individuals. At the same time, issues of transparency, access, and accuracy must not be neglected. Tene and Polonetsky (2013) discussed the distinction between identifiable and nonidentifiable data and considered de-identification methods (anonymization, pseudonymization, encryption, key coding) as important measures in data protection and security.

Schwartz (2011) developed a set of ethical principles for data analytics solutions based on a series of interviews with data privacy experts, lawmakers, and analysts. These principles include a set of overarching ethical standards related to compliance with legal requirements, compliance with cultural and social norms, accountability of measures tailored to identified risks, inclusion of appropriate safeguards to protect the security of data, and setting of responsible limits on analytics in sensitive areas or with vulnerable groups. In addition to setting up these generic principles, Schwartz (2011) argued that ethical principles must be adjusted to the various sub-stages of the analytics process, so the rules for how to tackle these challenges should be tailored to each stage of the analytics while also seeking to maximize good results and minimize bad ones for data subjects. Thus, in the data collection stage, care should be taken in regard to the type and nature of the information acquired, particularly in terms of avoiding the collection of sensitive data, while in the data integration and analysis stage, sufficient data quality must be ensured and anonymization should be performed. Finally, in the decision-making stage, the analytics results on which decisions are based must be ensured to be reasonably accurate and valid.

3.3 General Ethical and Privacy Guidelines

The OECD provides comprehensive guidelines and checklists for those who are seeking guidance on how to deal with privacy issues in analytics technologies and other systems (Spiekerman and Cranor 2009; Tene and Polonetsky 2013). In 1980, the OECD provided the first internationally agreed collection of privacy principles that harmonized legislation on privacy and facilitated the international flow of data

(OECD 2013b). The set of eight basic guidelines mirrored the principles earlier defined by the European Convention for the Protection of Individuals with Regard to the Automatic Processing of Personal Data (Levin and Nicholson 2005). The OECD principles are summarized as follows (OECD 2013b, pp. 14–15).

Individual Participation Individuals should have the right to obtain confirmation concerning whether data that is related to them is held and to have this data communicated to them, to be given reasons for a request denied, to challenge data that is related to them, and to have the data erased, rectified, completed, or amended.

Collection Limitation There should be limits to the collection of personal data. Data must be obtained by lawful and fair means and, where appropriate, with the knowledge or consent of the data subject.

Data Quality Personal data should be **relevant** to the purposes for which they are to be used and collected only to the extent necessary for those purposes. Data should be accurate and complete and kept up-to-date.

Accountability A data controller should be assigned who is accountable for complying with measures and regulations regarding ethics, privacy, and data protection.

Security Personal data should be protected by reasonable and sufficient security against loss or unauthorized access, destruction, use, modification, and disclosure.

Purpose Specification The purposes for which personal data are collected should be specified no later than at the time of data collection. Subsequent use should be limited to the fulfilment of those purposes or compatible purposes.

Openness There should be a general policy of openness about developments, practices, and policies with respect to personal data. Information on the existence and nature of personal data, the purpose for their use, and the identity and location of the data controller should be available.

Use Limitation Personal data should not be disclosed, made available, or used for purposes other than those specified except with the consent of the data subject or by the authority of the law.

Although these principles are not legally binding for OECD members, they have served as the basis for European legislation this this field (cf. Levin and Nicholson 2005; Spiekerman and Cranor 2009). The Guidelines on the Protection of Privacy and Transborder Flows of Personal Data (OECD 2013b), an update of the original version from 1980, maintains the original principles while considering cross-country data flows and strengthening privacy enforcement. The updated principles focus on the practical implementation of privacy protection in an increasingly digitized and "border-less" world and provide a risk-management type of approach.

The changing digital markets have made it necessary to consider the specificities of "big data." In a report on the broader topic of "data-driven innovation as a new source of growth," the OECD (2013a) mentioned data-intensive sectors and disciplines like online advertising, health care, logistics and transport, and public administration. However, the field of education was not addressed in this report.

Based on the framework of the (original) OECD guidelines, the Federal Trade Commission of the United States (1998) defined the Fair Information Practice Principles (FIPP), which specifies and exemplifies fair practices in the digital world. These practices can be summarized in terms of five main principles:

Notice and Awareness Data subjects shall be informed before their personal data is recorded to allow them to make conscious and informed decisions about whether they are willing to disclose their personal information and whether they approve the recording, processing, and use of their data. This principle is considered the most fundamental and as a prerequisite for the following principles.

Choice and Consent People should be able to make decisions about how their data is used by means of opt-in questions (explicitly giving consent to the data use) or opt-out questions (allowing data to be used by default unless the data subject refuses consent).

Access and Participation People must have the ability to access and control the data that have been recorded and to demand deletion or correction of data.

Integrity and Security All data are must be stored, processed, and archived in such a way that unwanted and possibly harmful data manipulation and unwanted access to data are avoided.

Enforcement and Redress Legislation that creates private remedies for users, and government enforcement is required to ensure compliance with privacy protection principles, enforcement and redress mechanisms through self-regulatory regimes.

These principles reflect attempts to cope with the perils of a digitized world and have been extended over time to meet new challenges. However, the focus has been on fields like health care and e-governance, not on education, although the focus could change with the increasing use of digital learning solutions, the advent of smart learning systems, and, not least, with the increasing use of learning analytics and educational data mining. Slade and Prinsloo (2013) argued that ethics and privacy issues in education are linked to general research ethics for the Internet, which is an important argument since the aforementioned principles arise primarily from an applied, market-oriented perspective. The Association of Internet Researchers has provided a set of ethical guidelines for decision-making (Markham and Buchanan 2012) that give researchers a code for conducting research in an ethical and professional manner. Because of such initiatives, the young field of learning analytics has emphasized the aspects of ethics and data protection, so a number of frameworks for ethics and data protection in learning analytics have been developed. The next section lists the most important and widespread of these frameworks.

3.4   Ethical Frameworks in Learning Analytics

Learning analytics is a field in which ethical issues, privacy, and data protection play critical roles, specifically when children are concerned. Researchers and developers in this field have held lively discussions about such issues and principles, so awareness of those aspects of learning analytics in the academic communities of learning analytics and educational data mining research is high. Although many authors have mentioned ethical issues and although there is a common understanding that such aspects of learning analytics are critical, only a few coherent approaches have elaborated on ethical challenges in any detail or attempted to define a framework to guide institutions, researchers, and developers in the application of learning analytics (Slade and Prinsloo 2013). This chapter presents the main ethical aspects and existing frameworks that can be considered as good guidelines.

The topics of privacy and ethics are also directly related to aspects of trust and accountability (Pardo and Siemens 2014), as students' trust in teachers allows them to invent, adapt, and excel. However, accountability constrains and confines teachers' tendency to experiment and encourages them to play it safe and conceal their mistakes (Didau 2015). The same is true for an analytics system, as the perceived trustworthiness of a learning analytics system determines its power and its impact; that is, analytics is only as good as the user believes it to be. At the same time, accountability limits the degree to which invalidated analytics are brought into practice.

Reflection and deliberation on ethical questions should be aligned with technical innovation in analytics because the slow pace of law may not match the speed of innovation. Nevertheless, existing approaches to ethics in learning analytics commonly and understandably ground their discussions on the legal understandings of privacy (Willis 2014). However, the legal view must be complemented by the more philosophical angle of trust and accountability.

One possible approach to elaborating these issues of learning analytics is to determine, analyze, and manage the risks of implementing a learning analytics solution. In their generic framework, Greller and Drachsler (2012) considered both the ethical and the legal aspects of learning analytics under the heading of external constraints. Apart from ethical, legal, and social constraints, they also considered organizational, managerial, and process constraints as components of this dimension. These external limitations can be categorized into conventions (ethics, personal privacy, and other socially motivated constraints) and norms (restrictions by law or mandated standards and policies), thus fortifying the argument that there are both a reasonable distinction and a close link between ethics and legal regulations. Ethics deals with what is morally allowed, while the law defines what is allowed without incurring legal consequences (Berg 2013).

Of course, in many cases, ethical issues are reflected in legislation, but ethical considerations go beyond what is set in law to depend on ideological assumptions and epistemologies (Slade and Prinsloo 2013). In addition, the appraisal of ethical viewpoints is to a large degree determined by culture such that different regions of the world differ in what is ethical conduct. Such is the case even in similar regions, such as the US and Europe. Because most legal regulations are based on ethics, an ethical position should be applied when interpreting the law (Sclater 2014b). Kay et al. (2012, p. 20) highlighted that, given the mission and responsibilities of education, "broad ethical considerations are crucial regardless of the compulsion in law," and explained that learning analytics implies conflict in ensuring educational benefits, meeting business interests, putting competitive pressure on educational institutions, and meeting the expectations of born-digital generations of learners. They postulate four key principles for good practice with respect to ethical aspects and analytics that account for these conflicts:

Clarity Clearly defining the purpose, scope, and limitations of analytics.

Comfort and Care Considering the interests and feelings of the data subject.

Choice and Consent Informing subjects and providing the option to opt out.

Consequence and Complaint Acknowledging the possibility of unforeseen consequences and providing mechanisms for complaint.

Willis et al. (2013) suggested using the so-called Potter Box, a flexible and generic ethical framework commonly applied in business communications to address ethical issues and dilemmas in learning analytics (Willis et al. 2013). They referred to the need to balance faculty expectations, privacy legislation, and an educational institution's philosophy of student development when dealing with ethical questions. The Potter Box provides a scaffolding for thinking and a framework for analyzing a situation but does not provide one clear solution to ethical dilemmas. The box contains four steps to take when making ethical decisions, as described in Table 3.1.

Table 3.1 The Potter Box

Definition: The empirical facts of a given situation are clearly defined, without judgment.	**Loyalties**: Loyalties are chosen: for example, people who are affected by a situation (application of learning analytics) entities that will act on the information gained, and persons responsible in case of failure, etc.
Values: Values that represent conventions, rights, and beliefs (e.g., moral values, professional values) are identified and compared. Differences in the perspectives of the stakeholders involved can be analyzed.	**Principles**: A set of ethical principles that are applicable to the situation is identified (e.g., Mill's principle of utility: "Seek the greatest happiness for the greatest number").

Slade and Prinsloo (2013) took a socio-critical perspective on the use of learning analytics and proposed six principles to address ethics and privacy challenges:

Learning Analytics as a Moral Practice Key variables are effectiveness, appropriateness, and practical necessity, and the goal is to understand, rather than to measure.

Students as Agents Students should be involved in the learning analytics process as collaborators and co-interpreters, so a student-centric approach to learning analytics is recommended.

Student Identity and Performance Are Temporal Dynamic Constructs Learning analytics often provides only a snapshot of a learner over a limited time span and in a particular context and do not reflect long-term information unless longitudinal data are used.

Student Success Is a Complex and Multidimensional Phenomenon Learning progress and learning success are determined by multidimensional, interdependent interactions and activities. The data used for learning analytics is always incomplete and may lead to misinterpretation or bias.

Transparency Information about the purpose of data use, data controllers/processors, and measures to protect the data should be disclosed in suitable forms.

Education Cannot Afford Not to Use Data Educators cannot ignore the information that learning analytics may provide if the best possible results for individual learners are to be reached.

Pardo and Siemens' (2014) analysis of solutions for privacy- and ethics-related issues in educational organizations resulted in a set of four principles that can serve as a basis for setting up appropriate mechanisms for meeting social, ethical, and legal requirements when developing and deploying learning analytics.

Transparency All stakeholder groups in learning analytics (e.g., learners, teachers, educational administrator) should receive information about what type of data is collected and how it is processed and stored.

Right to Access The security of data must be guaranteed and rights to access a dataset clearly defined.

Student Control Over Data Students shall be granted access rights to data to control and, if necessary, require a correction of data.

Accountability and Assessment The analytics process should be reviewed and the responsible entities identified for each aspect of the learning analytics scenario.

Another general approach is Sclater and Bailey's (2015) code of practice for learning analytics. The code of practice, which is based on an extensive literature review of legal and ethical issues in learning analytics (Sclater 2014a), addresses eight themes:

Responsibility Identifying who is responsible for the data and data processing for learning analytics in an institution.

Validity Ensuring the validity of algorithms, metrics, and processes.

Privacy Ensuring protection of individual rights and compliance with data protection legislation.

Transparency and Consent Ensuring openness on all aspects of using learning analytics and meaningful consent.

Minimizing Adverse Impacts Avoiding potential pitfalls for and harm to students.

Access Providing data subjects access to their data and analytics.

Enabling Positive Interventions Handling interventions based on analytics in a positive, appropriate, and goal-oriented way.

Stewardship of Data Handling, processing, and storing of data appropriately and with accountability.

One of the most important frameworks of privacy and ethics issues in learning analytics is the DELICATE checklist (Drachsler and Greller 2016). The checklist was developed after intensive studies of EU law and various expert workshops on ethics and privacy for learning analytics. DELICATE stands for

D-etermination Determine the purpose of learning analytics for your institution.

E-xplain Explain the scope of data collection and usage.

L-egitimate Explain how you operate within legitimate legal frameworks, referring to the essential legislation.

I-nvolve Involve stakeholders and give assurances about the data's distribution and use.

C-onsent Seek consent through clear consent questions.

A-nonymise Anonymize individual data as much as possible.

T-echnical Aspects Monitor who has access to data, especially in areas with high staff turnover.

E-xternal Partners Make sure external partners provide the highest data-security standards.

Learning analytics solutions that are specifically designed for the K-12 school context have been developed in the context of the European project called Lea's Box (http://www.learning-analytics-toolbox.org/tools/). The project emphasized the extreme diversity and heterogeneity of daily school practices in terms of multiple and overlapping learner groups, multiple and overlapping subjects, main lessons and afternoon care, multiple teachers in one subject area, long time periods to be covered, a generally technology-lean setting, and the use of all types of electronic tools, apps, and devices. The Lea's Box solutions build on bringing all the scattered data sources that can be used for learning analytics together and aggregating the information from those sources. Such an effort bears particular ethics and privacy risks, so during the project a comprehensive ethics and privacy framework was developed (Steiner et al. 2016). A particular emphasis of this framework lies on automatic analytics algorithms that aggregate and process the data and result in certain higher-level assessments and performance predictions that are far beyond descriptive statistics. Steiner et al. (2016) observed that the main algorithms that are already applied in practice did not undergo thorough validation studies, so the algorithms' results and predictions validity and reliability cannot be guaranteed. The framework includes eight aspects:

Data Privacy In principle, not all information that may be available in digital form should be accessible to teachers. One must consider in the conceptualization phase of a learning analytics system which data should be accessed and made part of analytics (e.g., discarded working steps outside the time when work assignments were submitted).

Purpose and Ownership In the conceptualization phase, the concrete purposes and goals of a learning analytics system must be made clear. A data-mining approach—that is, collecting all possible data and determining later on what can be done with it—is not acceptable. Who owns the data must also be precisely set out. For example, organizations may own final aggregations and analytics (e.g., grades), while the individual learners still own all the raw data that led to these results.

Consent While consent is an important good from an ethics point of view, consent is not always possible in school contexts, where the ability to opt out is not practical. Broad consent of the majority of students or their legal representatives can facilitate the establishment of a common set of organizational practices.

Transparency and Trust Stakeholders' trust in the system's functions and outcomes is the most valuable asset for a learning analytics system that evaluates performance. A learning analytics solution should make its principles and analytics algorithms appropriately transparent to all stakeholders, and learners should have the option to disagree with the results of analytics and the opportunity to negotiate results. (Such is the case in persuadable open learner modelling systems.)

Access and Control Related to the aspect of transparency, access and control can strengthen the trust in a learning analytics system. Specifically, learners should have the ability to access the information gathered about them easily and to correct the data if necessary.

Accountability and Assessment Organizations that set up learning analytics solutions should be accountable, so who is responsible for what should be clear and public. The functions, the validity, and the impact of the results should be evaluated frequently so erroneous analytics and misuse of outcomes can be identified.

Data Quality Learning analytics is based solely on data from various resources, and these data are usually incorrect and incomplete to a degree. In the context of the conceptualization phase and in the course of maintaining a learning analytics system, the data's quality and (sufficient) completeness must be evaluated and revisited frequently with emphasis on the question concerning whether the available data serve the learning analytics' stated goals.

Data Management and Security   Data protection regulations and the principles of privacy make clear demands on how data is treated, including which data to store (e.g., raw data vs. aggregated results), in which form, where, and for how long. In the conceptualization phase of a learning analytics solution, these issues must be considered and solutions implemented (e.g., that certain datasets are deleted automatically after a given time). Conceptual and technical solutions must also address what happens when students enter or leave the institution—whether to bring in new data for the new students and whether to take out data from the departed students. Finally, the technical solutions must have an up-to-date level of security.

3.5   Ethics by Design

The principles of ethics, security, privacy, and data protection influence the practice of learning analytics solution, while the possibilities and functions of a proper realization of the core principles should already be part of the technology itself. Therefore, the principles should be considered in the design phase or when choosing a learning analytics solution. The integration of ethics in the conceptualization and development phase is usually referred to as "privacy by design," "value-sensitive design," or "ethics by design" (e.g., Bomas 2014; Scheffel et al. 2014).

Value-sensitive design and ethics by design mean seamlessly integrating ethical and legal requirements and practices into the design, development, and evaluation process of a software (Friedman 1997) so the software itself follows ethical rules and supports its users in following ethical rules (Gotterbarn 1999). Privacy by design, on the other hand, focuses more concretely on privacy engineering and developing guidelines for designing privacy-friendly systems (Cavoukian 2011). Spiekerman and Cranor (2009) provided an analysis of privacy requirements that can be applied to a variety of software systems. Their analysis identified activities that information systems typically perform and highlighted the impact of these activities on users' privacy, depending on how the system performs the activities, the type of data that is used, the stakeholders' characteristics, and the aspects of privacy that are affected. Spiekerman and Cranor (2009) provided guidelines on how notice, choice, and access can be implemented as fair information practices and how users can be informed about them. These authors also defined a "privacy-by-policy" and a "privacy-by-architecture" approach that focus on implementation of notice and choice principles, minimizing the collection of identifiable personal data, and anonymization.

3.6 Important Legal Regulations

The privacy and data protections that are regulated by national and international law address the disclosure or misuse of private individuals' information. It is out of the scope of the book to provide a complete overview, but here are some main examples of laws and policies that also affect learning analytics.

Regulations for online systems were driven by countries with high Internet usage (Pardo and Siemens 2014). Examples are the European Union Directive on the protection of individuals with regard to processing of personal data and the free movement of such data (European Parliament 1995), the Canadian Personal Information Protection and Electronic Documents Act (2000), the Australian Privacy Act and Regulation (1988, 2013), and the US Consumer Data Privacy in a Networked World (The White House 2012). The Family Educational Rights and Privacy Act or FERPA (2004), a US federal law that applies specifically to the privacy of students' education records, allows the use of data on a need-to-know basis and provides parents with certain rights of access to their children's education records. In parallel with legislative efforts to protect data, nonprofit organizations have evolved that defend users; digital rights, including the Electronic Frontier Foundation and Privacy Rights Clearinghouse in the United States.

The European legislation plays a special role since it deals with the necessary interplay of national, EU-based, and international regulations. The transfer of personal data between countries in the EU is necessary in companies' and public authorities' day-to-day business. Since conflicting data-protection regulations might complicate international data exchanges, the EU has established common rules for data protection, the application of which in each EU country is monitored by national supervisory authorities.

The European data protection legislation considers the protection of personal data a fundamental right. The current EU law is the 2018 General Data Protection Directive, which applies to countries of the European Economic Area (EEA) (i.e., all EU countries plus Iceland, Liechtenstein, and Norway). The directive seeks a high level of protection of individual privacy and to control the movement of personal data within the European Union, whether the data is collected and processed automatically (in a digital form) or in non-automated ways (traditional paper files). Each member state is to apply the provisions nationally (cf. Steiner et al. 2016). The main purpose of the GDPR is to protect all EU citizens from privacy and data breaches in an increasingly data-driven world.

In the context of schools and small- to medium-scale universities, data protection might have played a subordinate role in the past, but this must change. If you want to read more about data protection and ethics, in http://www.learning-analytics-tool box.org/data-protection/ you can find Kickmeier-Rust and Steiner's (2018) summary of key aspects and advice for how to address the requirements in a practical way.

3.7   Conclusions

Moore (2008) highlighted ethics as a critical aspect of learning analytics, but ethical considerations are full of contradictory viewpoints and squishy definitions, as individual beliefs, values, and preferences influence the scientific work in the contexts of designing, developing, and deploying educational software. It is inevitable that data protection, data privacy, and ethical considerations will soon be a more regular part of education. The fulfillment of these requirements is not always easy, especially in typical schools or small- to medium-scale institutions of higher education.

This chapter provided an overview of general regulations and possible solutions in the form of ethics frameworks that are intended to provide the foundations for a proper and practical code of conduct. It also provided a framework to serve as a scaffolding for planning, developing, installing, and applying learning analytics in an organization's daily practice. For example, it is highly recommended to use a checklist like the DELICATE when implementing learning analytics solutions. The technology and tools that are applied to analyzing and reasoning about learners' data should be in line with these foundations.

The regulations that protect data and privacy should not stop educators from benefitting their students and curricula by using digital data and analytics solutions so long as educators keep in mind the need to protect students' data in general and individual privacy in particular. Education and training cannot afford not to use innovative technologies for the benefit of their learners and teachers (Slade and Prinsloo 2013).

References

Alexander, A. (2015). Digital surveillance 'worse than Orwell', says new UN privacy chief, *The Guardian*. Retrieved from https://www.theguardian.com/world/2015/aug/24/we-need-geneva-convention-for-the-internet-says-new-un-privacy-chief

Anderson, T. D., & Gardiner, G. (2014). What price privacy in a data-intensive world? In M. Kindling & E. Greifeneder (Eds.), *Culture, context, computing: Proceedings of iConference 2014* (pp. 1227–1230). Illinois: iSchools. Retrieved from https://www.ideals.illinois.edu/handle/2142/47417

Barnes, S. B. (2006). A privacy paradox: Social networking in the United States. *First Monday, 11* (9). Retrieved from http://firstmonday.org/article/view/1394/1312

Berg, A. (2013). *Towards a uniform code of ethics and practices for learning analytics* [Web log post, August 21]. Retrieved from https://www.surfspace.nl/artikel/1311-towards-a-uniform-code-ofethics-and-practices-for-learning-analytics/

Bomas, E. (2014). *How to give students control of their data* [Web log post, August 29]. Retrieved from http://www.laceproject.eu/blog/give-students-control-data/

Cavoukian, A. (2011). *Privacy by design: The 7 foundational principles. Implementation and mapping of fair information practices.* Ontario, Canada: Information and Privacy Commissioner of Ontario. Retrieved from https://www.ipc.on.ca/wp-content/uploads/Resources/7foundationalprinciples.pdf

Debatin, B., Lovejoy, J. P., Horn, A.-K., & Hughes, B. N. (2009). Facebook and online privacy: Attitudes, behaviors, and unintended consequences. *Journal of Computer-Mediated Communications, 15*(1), 83–108.

Didau, D. (2015). *Trust, accountability and why we need them both*. Retrieved from https://learningspy.co.uk/leadership/trust-vs-accountability/

Drachsler, H., & Greller, W. (2016). Privacy and analytics: It's a DELICATE issue a checklist for trusted learning analytics. In *Proceedings of the sixth international conference on Learning Analytics & Knowledge (LAK '16)* (pp. 89–98). New York: ACM.

European Parliament. (1995). *Directive 95/46/EC*. Retrieved from https://eur-lex.europa.eu/legal-content/EN/TXT/?uri=celex:31995L0046

Family Education Rights and Privacy Act (FERPA) 34 C.F.R.§ 99.34. (2004). Retrieved from https://www.law.cornell.edu/cfr/text/34/part-99

Federal Trade Commission. (1998). *Privacy online: A report to Congress*. Retrieved from https://www.ftc.gov/sites/default/files/documents/reports/privacy-online-report-congress/priv-23a.pdf

Ferguson, R. (2012). Learning analytics: Drivers, developments and challenges. *International Journal of Technology Enhanced Learning, 4*(5-6), 304–317.

Friedman, B. (1997). *Human values and the design of computer technology*. Cambridge, MA: Cambridge University Press.

Gotterbarn, D. (1999). How the new software engineering code of ethics affects you. *IEEE Software, 16*(6), 58–64.

Greller, W., & Drachsler, H. (2012). Translating learning into numbers: A generic framework for learning analytics. *Educational Technology & Society, 15*(3), 42–57.

Kay, D., Korn, N., & Oppenheim, C. (2012, November). Legal, risk and ethical aspects of analytics in higher education. *JISC CETIS Analytics Series, 1*(6). Retrieved from http://publications.cetis.org.uk/wp-content/uploads/2012/11/Legal-Risk-and-Ethical-Aspects-of-Analytics-in-Higher-Education-Vol1-No6.pdf

Kickmeier-Rust, M.D., & Steiner, C.M. (2018). *Data protection and ethics in learning analytics. LA4S white paper*. Retrieved from http://la4s.iwiserver.com/wp-content/uploads/sites/42/2018/06/DataProtectionWhitePaper.pdf

Levin, A., & Nicholson, M. J. (2005). Privacy law in the United States, the EU and Canada: The allure of the middle ground. *University of Ottawa Law & Technology Journal, 2*(2), 357–395.

Markham, A., & Buchanan, E. (2012). *Ethical decision-making and Internet research: Recommendations from the AoIR Ethics Working Committee (Version 2.0)*. Retrieved from http://aoir.org/reports/ethics2.pdf

Merriam Webster Dictionary. (2019). Retrieved from http://www.merriam-webster.com/

Moore, S. L. (Ed.). (2008). *Practical approaches to ethics for colleges and universities*. San Francisco, CA: Jossey-Bass.

OECD. (2013a). Exploring data-driven innovation as a new source of growth: Mapping the policy issues raised by "big data." In *Supporting investment in knowledge capital, growth and innovation*. Paris: OECD Publishing.

OECD. (2013b). *The OECD privacy framework*. OECD Publishing. Retrieved from http://www.oecd.org/sti/ieconomy/oecd_privacy_framework.pdf

Pardo, A., & Siemens, G. (2014). Ethical and privacy principles for learning analytics. *British Journal of Educational Technology, 45*(3), 438–450.

Personal Information Protection and Electronic Documents Act. (2000, c.5). Retrieved from the Department of Justice Canada website http://laws-lois.justice.gc.ca/eng/acts/P-8.6/

Privacy Act 1988. No. 119 1988. (Cth). Retrieved from the Australian Federal Register of Legislation website https://www.legislation.gov.au/Details/C2012C00414

Privacy Regulation 2013 (Cth). Retrieved from the Office of the Australian Information Commissioner website https://www.oaic.gov.au/privacy-law/privacy-act/privacy-regulations

Richards, N. M., & King, J. H. (2014). Big data ethics. *Wake Forest Law Review, 49*, 393–432.

Scheffel, M., Drachsler, H., Stoyanov, S., & Specht, M. (2014). Quality indicators for learning analytics. *Educational Technology & Society, 17*(4), 117–132.

Schwartz, P. M. (2011). Privacy, ethics, and analytics. *IEEE Security and Privacy, 9*(3), 66–69.

Sclater, N. (2014a). Code of practice for learning analytics: A literature review of the ethical and legal issues. *JISC*. Retrieved from https://repository.jisc.ac.uk/5661/1/Learning_Analytics_A-_Literature_Review.pdf

Sclater, N. (2014b, October 29). *Notes from Utrecht workshop on ethics and privacy issues in the application of learning analytics*. Effective Learning Analytics. Jisc. [Web log post]. Retrieved from   https://analytics.jiscinvolve.org/wp/2014/10/29/notes-from-utrecht-workshop-on-ethics-and-privacy-issues-in-the-application-of-learning-analytics/

Sclater, N., & Bailey, P. (2015). *Code of practice for learning analytics*. Jisc. Retrieved from https://www.jisc.ac.uk/guides/code-of-practice-for-learning-analytics

Slade, S., & Prinsloo, P. (2013). Learning analytics: Ethical issues and dilemmas. *American Behavioral Scientist, 57*(10), 1509–1528.

Spiekerman, S., & Cranor, L. F. (2009). Engineering privacy. *IEEE Transactions on Software Engineering, 35*(1), 67–82.

Steiner, C. M., Kickmeier-Rust, M. D., & Albert, D. (2016). LEA in private: A privacy and data protection framework for a learning analytics toolbox. *Journal of Learning Analytics, 3*(1), 66–90.

Stutzman, F., & Kramer-Duffield, J. (2010). Friends only: Examining a privacy-enhancing behavior in Facebook. In *Proceedings of the SIGCHI conference on human factors in computing systems* (pp. 1553–1562). New York: ACM.

Tene, O., & Polonetsky, J. (2013). Big data for all: Privacy and user control in the age of analytics. *Northwestern Journal of Technology and Intellectual Property, 11*(5), 239–273.

The White House. (2012). *Consumer data privacy in a networked world: A framework for protecting privacy and promoting innovation in the global digital economy*. Washington, DC: The White House. Retrieved from http://cdm16064.contentdm.oclc.org/cdm/ref/collection/p266901coll4/id/3950

Willis, J. E. (2014). Learning analytics and ethics: A framework beyond utilitarianism. *EDUCAUSE Review Online*. Retrieved from https://er.educause.edu/articles/2014/8/learning-analytics-and-ethics-a-framework-beyond-utilitarianism

Willis, J. E., Campbell, J. P., & Pistilli, M. D. (2013). Ethics, big data, and analytics: A model for application. *EDUCAUSE Review Online*. Retrieved from http://er.educause.edu/articles/2013/5/ethics-big-data-and-analytics-a-model-for-application

Part II
Selected Learning Analytics Recipes

Chapter 4
Understanding Student-Driven Learning Processes

Abstract Nowadays, teachers often use digital platforms to support their face-to-face activities in the classroom. This can spice up the teaching or facilitate learning activities from home. After using such platforms, it could be interesting and beneficial to visualize the students' actions to better understand the patterns in the students' learning and performance. Often the problem with using existing learning platforms is that the visualizations are pre-defined by the system designer and might not fit well for the purpose and objectives of the teacher or for a specific course. In this recipe we tackle this lack of flexibility in creating visualizations by introducing DojoIBL. We showcase an example of how dashboards and own visualizations of students' digital traces can be created. Teachers without previous experience can also easily do this on their own. In addition, teachers can also use DojoIBL to design and create courses, projects or seminars.

Keywords DojoIBL · Project management tool · Inquiry-based learning

4.1 Appetizer

Policymakers recognize student-centered approaches as efficient ways to make students more proficient in their learning. However, such student-centered approaches require students to take ownership of their learning, so students must develop skills like reflection, self-regulation, metacognition, and self-assessment that help them to be autonomous, while teachers need to follow students' progress and to intervene in their learning when needed. These two aspects of a student-centered approach can be facilitated with learning analytics.

One of the problems that teachers face when using learning analytics in an educational platform—if one even exists—is the visualizations' lack of flexibility. Usually, they are fixed and given by the system designers, so they may not fit the learning purpose or the teacher's objectives. Therefore, teachers might not find learning analytics the perfect instrument for helping them monitor and support their students.

R. Jaakonmäki et al., *Learning Analytics Cookbook*, SpringerBriefs in Business Process Management, https://doi.org/10.1007/978-3-030-43377-2_4

We address this issue in this recipe by introducing DojoIBL (Suárez et al. 2016, 2017) in the preparation section and then showing in the cooking section how teachers can easily create their own visualizations in the system. Teachers can create courses, projects, or seminars, and the students can follow them in a structured way. The recipe ends with a concrete example of how a teacher with no previous experience with the system can easily create a visualization for his or her specific scenario. We promise that, in less than an hour, you could be using DojoIBL with your students and using customized learning analytics dashboards to monitor them.

Preparation Time: 1 h

4.2   Ingredients

This recipe offers you a full learning analytics menu that can be shared with your students using the following basic ingredients:

- DojoIBL platform with Stefla dashboard http://www.learning-analytics-toolbox. org/tools/, which is free, always online, and ready to use
- An hour of your time

4.3   Preparation

The first step of the recipe is at the DojoIBL platform. Briefly, DojoIBL is a lightweight project-management tool especially developed for educators. In this section we explain its basic functionality that will allow you to create a project and invite your students to the system in less than 5 min.

Before we jump into the details, open the following link and click the DojoIBL "View Product" button to start this tasty recipe: http://www.learning-analytics-toolbox.org/tools/.

Voila! Next to the logo, you see two ways to access DojoIBL: (1) Creating your own DojoIBL account with a valid email, which allows you to stay away from third-party log-in systems like Google, which might be problematic for some schools, or (2) Using the Google system for authentication. If you want to have immediate access to DojoIBL, this approach will give you access in less than a minute.

⌛*Cooking pause: As in every recipe, patience is essential, so I will wait for you to log in using one of the options presented above.*

Once you've logged in, you should see an empty screen in which your projects will show up once we have created them. Click the "+ Create new project" button, and you will see a catalogue of project templates that have been extracted from the literature. Creating a project from scratch requires a lot of design time, so the site

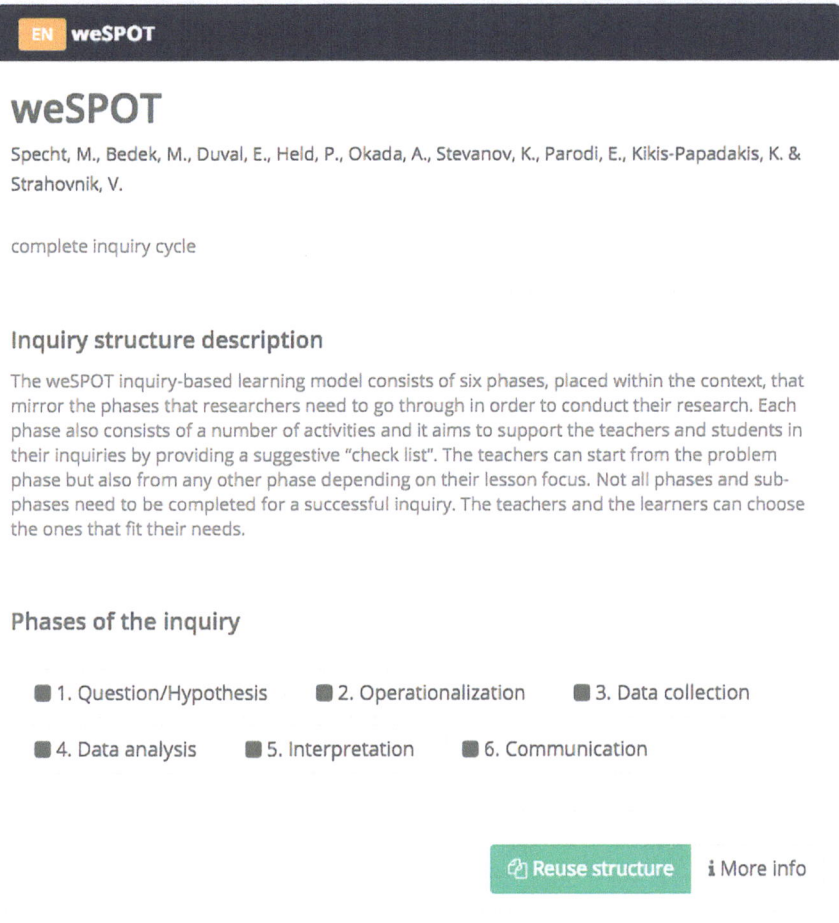

Fig. 4.1 Example of a reusable project template for DojoIBL

provides the flexibility to create your own project structures and customizable templates that can inspire you. Some of these templates are specific to inquiry-based learning, some are meant to increase metacognitive awareness, and others, like the one in Fig. 4.1, are pedagogical models developed in the European weSPOT Project. Any of the available templates can be used for the purpose of this recipe. It's up to you to decide.

Having templates with your favorite project structures can really speed up the management of your classroom, course, or seminar project. When you have decided on your template, click the "Reuse structure" button.

⌔*Cooking pause: Before continuing, be sure you have created (or reused) a structure for your project.*

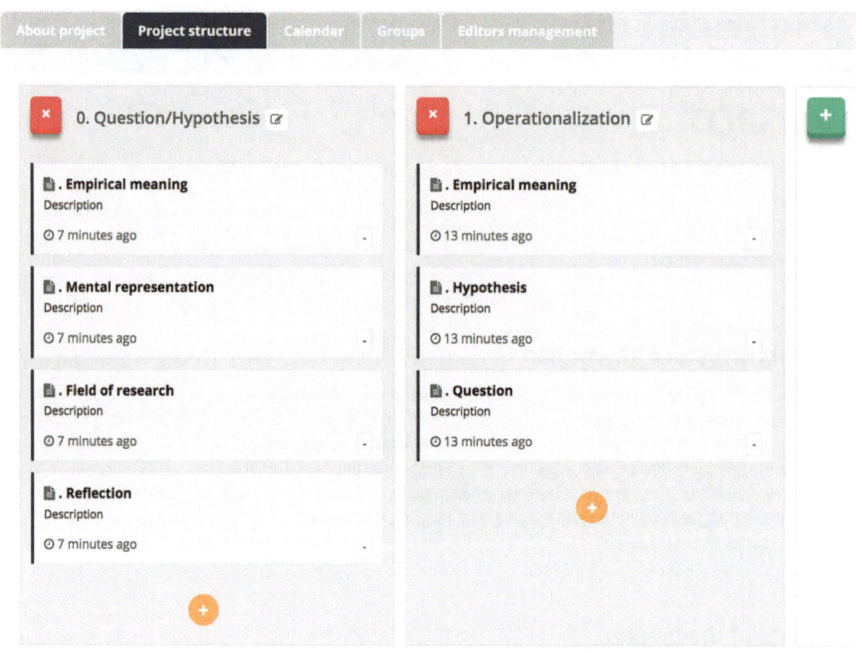

Fig. 4.2 Screenshot from a research project seminar structure in DojoIBL

By now, you should be in the editor mode, which is a dashboard for managing the project you have just created. The dashboard presents five tabs that contain information on the project. The one that matters now is "project structure," which contains the project's activities, organized by phases, as shown in Fig. 4.2.

In this interface, there are three main functionalities that allow you to modify the structure as you wish. To add new phases to the project, use the [**+ Green Button**]. To add activities to a phase, use the yellow [+ *Yellow Button*]. To remove a phase from the inquiry structure, use the red [***x Red Button***]. To modify one of the activities, just click on its title.

Great! You have created your first project with DojoIBL, but where are the students? How can you add them to the project?

In DojoIBL we have project structures and groups. The groups are instances of the project structures that the students use. For instance, you can have your classroom divided in many groups working under the same structure but independent from each other. To create them, go to the tab groups and use the [+ Add] button at the top right. Type the name of the group, click [+ Add], and the new group will be generated. Each group comes with a code that can be used for students to join. In Fig. 4.3, each colored circle corresponds to a user who has joined the group.

Once the group is created, you can access the students' view using the "view" button on the right side. You will see a list of phases in sequence and the activities

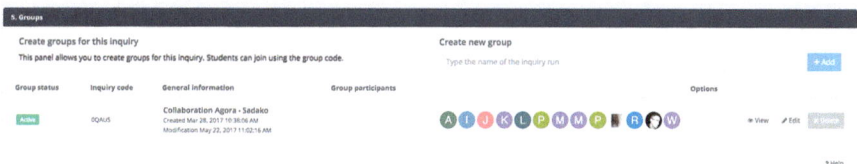

Fig. 4.3 An overview of the groups

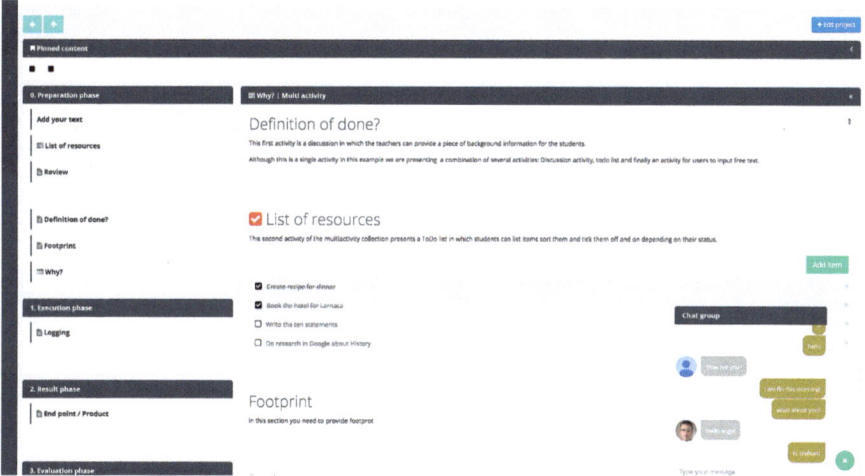

Fig. 4.4 Activity screen

underneath. If you click on one of the activities, you will see the visualization shown in Fig. 4.4.

The second step of our recipe is on the learning analytics dashboard platform called StefLa (Fig. 4.5). StefLa is a flexible dashboard that allows you to manage your queries (on the data from your platform) and the visualization widgets that will be embedded in your platform later.

🖐 ***Warning:*** *This section is a bit technical. If you are not interested in technical details and you just want to use DojoIBL with some standard visualizations, you can jump directly to the next cooking pause.*

The next screen allows designers to create queries about the data that is collected from the platform (DojoIBL in our example).

The collection of data from the platform is made through a RESTful API using a POST method to the following endpoint: /data-proxy/xAPI/statement/origin/ DojoIBL. Once the connection is established between the platform (DojoIBL) and the StefLa, you need to create a query to obtain the data for the later visualizations. In this example, we would like to see how many contributions the students have made to the activities provided in our platform.

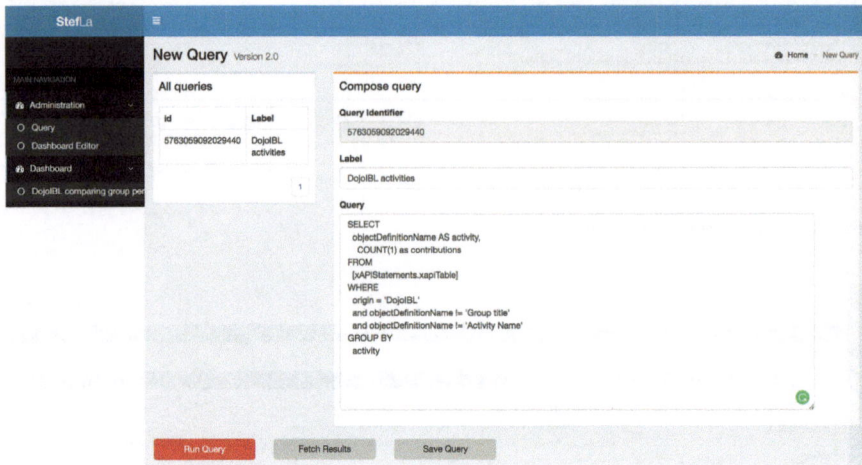

Fig. 4.5 Screenshot from StefLa that helps to create queries and visualization widgets

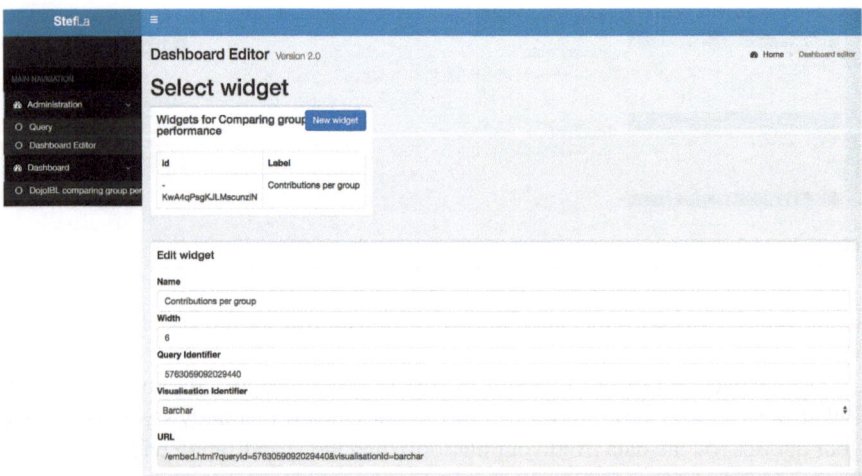

Fig. 4.6 Selecting a widget in StefLa Dashboard Editor

From the query we created, we will have some data that will be used in the visualizations, so now we need to create the visualization. The next screen shows how to do this (Fig. 4.6). The first block shows the list of visualizations, and the one below contains the details for each of the visualizations/widgets.

In the example, the name of the visualization is "Contributions per group." Each visualization has an identifier and the option to select the type of visualization. In this example we have gone for a "Barchar," which provides a bar chart that displays categorical data using rectangular bars.

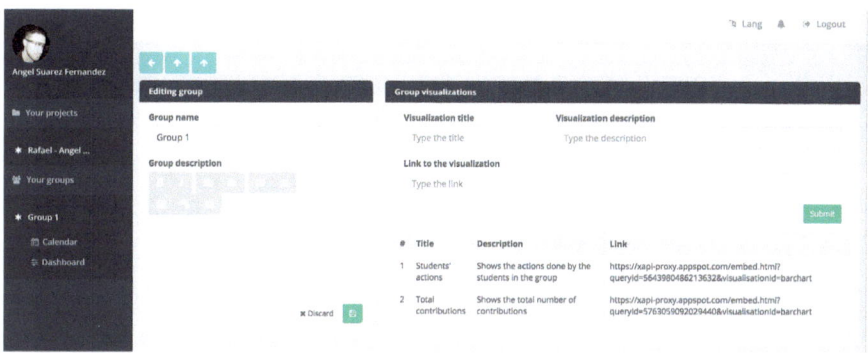

Fig. 4.7 Editing group visualizations

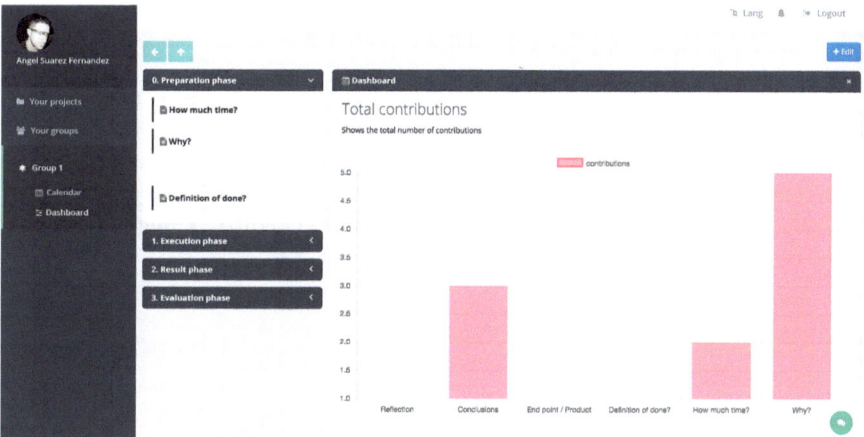

Fig. 4.8 Number of contributions

🍳*Cooking pause: Okay, we're done with the technical stuff. Just copy the URL from Stefla (Fig. 4.6) and paste it into DojoIBL.*

Let's assume at this point that the visualization is ready and we want to integrate it into DojoIBL. For that we will copy the link and paste it into the next section in DojoIBL. This section can be found in the "Group visualizations" of a project because each group should be able to have its own visualizations.

If you used Stefla analytics dashboard for creating a visualization, on the right side of the interface you can add the title, the description and the link to the visualization that was provided by the StefLa (Fig. 4.7).

Back on the student's view, you can find the visualization under the dashboard link on the left menu. In the example in Fig. 4.8, the activities shown in the overview (left) are also displayed in the visualization with the number of contributions (right). If you added more visualizations from the StefLa, they will appear below the ones provided by DojoIBL.

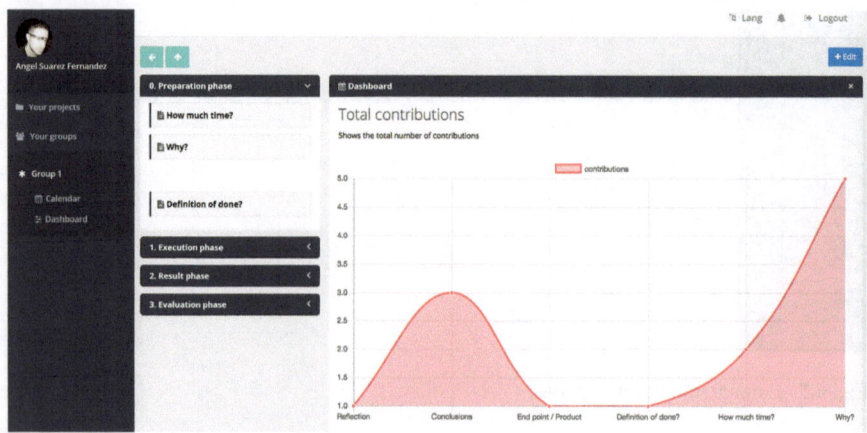

Fig. 4.9 Alternative way to visualize the number of contributions

Figure 4.9 shows another example of a visualization with the same query we used before so you can see that the system allows for multiple representations of the same data, and you can try out which way of presenting the data works best for you.

When you have reached this point in the recipe, you are able to integrate learning analytics visualizations from the StefLa platform into DojoIBL. **Congratulations**!

4.4   Cooking

1. Start with the DojoIBL structure and some students who are using it.
2. Add a query that retrieves the data you want to visualize.
3. Create a visualization and obtain the link to insert in DojoIBL.
4. Work in DojoIBL.
5. Look at the dashboard section in one of the groups to check the visualizations.

4.5   Serving

Let's work with a real example. Let's imagine that you are a Spanish teacher in a second-language-acquisition course, and you want to use inquiry-based learning with your students. You are planning to use DojoIBL to give structure to the inquiry process and to provide an effective way to support students' inquiry process.

First, think about what you want your students to investigate. In this example, you could ask your students to research how satirical magazines use humor and irony to convey critical messages to political parties, religions, and public figures.

In an inquiry project, the teacher usually sets the theme of the investigation, but the students are in the driver's seat, managing their own learning and exploring what

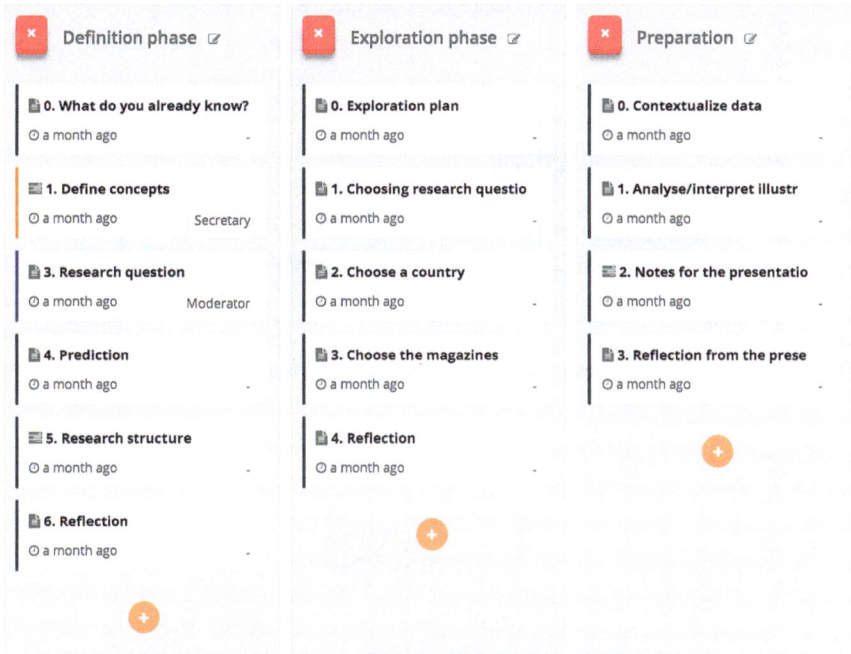

Fig. 4.10 A structure view of the phases of inquiry-based learning

they are interested in. However, in this example, we assume the students do not have a lot of practice in inquiry learning and that you have created the structure from beginning to end.

As Fig. 4.10 shows, the definition phase starts by asking students to define the inquiry's topic, concepts, and research questions. For this course, the second step is not useful so we exclude it from the definition phase. At the end of the first phase, students should come up with a prediction about what they want to learn and organize the research structure. In this activity, the students discuss whether the rest of the phases and activities are suitable for their investigation of the use of humor and irony in satirical magazines.

Once the structure is ready, the students start working on the topic, following the structure. In this example we have two groups, group A and group B.

Your role as a teacher is to provide guidance on the students' performance and the work they are doing. In this example, we monitor the students' activity using quantitative performance indicators, which give us information about the contributions the two groups make during the phases of the inquiry and during the activities in each phase. We use two visualizations to represent this information, one that shows the number of the students' interactions during the project phases, and the other that shows the amount of activities under a certain phase.

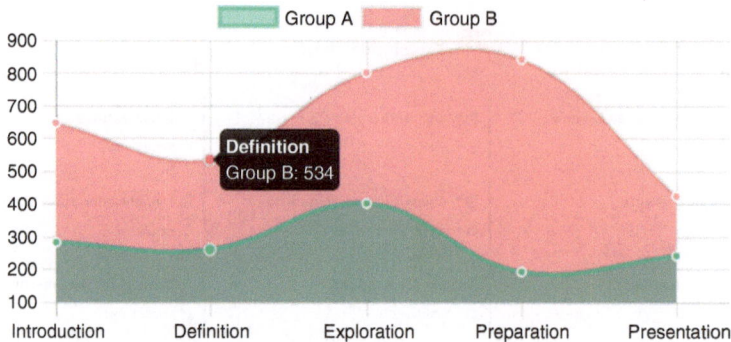

Fig. 4.11 Contributions throughout the phases of the inquiry project

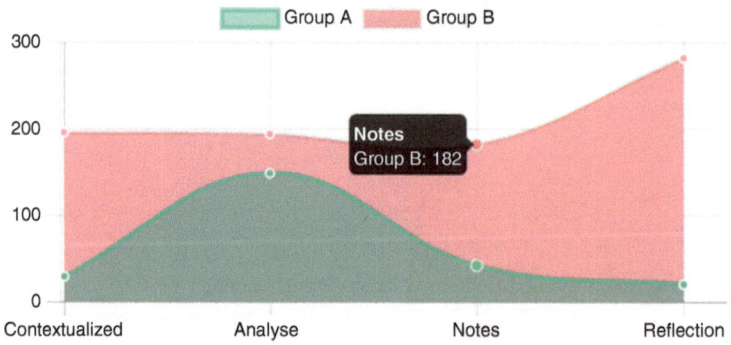

Fig. 4.12 Contributions in the preparation phase of the inquiry project

As shown in Fig. 4.11, group B contributed more to DojoIBL than group A did. As a teacher, you can use this information to help you judge the groups' performance. The visualization gives you several insights. First, group B contributed more, but both groups followed similar patterns throughout the phases, with the exception of the preparation phase. Second, in the preparation phase, something happened that caused group A to contribute less. You know your students, and at the end of the inquiry, you have an idea of which group obtained better results. If group A scored lower, one of the explanations could be in the preparation phase. Poor preparation might have affected their final result.

To look into what happened in the preparation phase, you can use the second visualization (Fig. 4.12), which shows each group's contributions to the various activities. As you can see, group B contributed more, particularly in the contextual and the reflection activities.

In this example two visualizations help you understand in which parts of the process students put their effort. The information is quantitative, but throughout the course of a long inquiry project, aggregating this type of information can improve your ability to evaluate your students' performance.

On a more general note, this recipe shows the flexibility of DojoIBL and a learning analytics dashboard that together help teachers like you visualize the information that is important to you. Aggregating data in this flexible way can benefit teachers by giving them insights into students' performance, especially in student-driven methodologies like inquiry-based learning.

References

Suárez, Á., Ternier, S., Prinsen, F., & Specht, M. (2016, September). DojoIBL: Nurturing communities of inquiry. In K. Verbert, M. Sharples, & T. Klobucar (Eds.), *11th European Conference on Technology Enhanced Learning (EC-TEL2016)* (pp. 533–536). Lyon: Springer.
Suárez, Á., Ternier, S., & Specht, M. (2017). DojoIBL: Online inquiry-based learning. In H. Xie, E. Popescu, G. Hancke, & M. B. Fernández (Eds.), *16th International Conference on Advances in Web-Based Learning – ICWL 2017*. Cape Town: Springer.

Chapter 5
Learning Analytics in a Primary School: Lea's Box Recipe

Abstract The amount of digital data about learning-related activities and achievements is steadily increasing. Yet, it is very difficult to combine all this data and to visualize learning patterns, even when data is readily available. The challenge is that data is gathered from various sources and might be stored in completely different formats. Therefore, educators cannot get comprehensive insights on the learning processes of their own students. To solve this problem, one would need to combine and aggregate all the available data and to link it to a suitable model of learning. Lea's Box addresses these issues and offers a central learning analytics platform that allows linking different data sources, so that teachers can do insightful analyses and reports. In addition, the system allows teachers to take notes on learning achievements and the learning progress of their students.

Keywords Lea's box · Learning analytics platform · Open learner modelling

5.1 Appetizer

Imagine you are a teacher in a nondigital or blended scenario like a typical school setting. Classroom teaching, where the teacher is lecturing in front of the class, the conventional blackboard, textbooks, and paper are the most common teaching activities and learning tools. Certainly the use of media like televisions, digital whiteboards, computers, tablets, and projectors is increasing in the classroom. Some progressive schools and institutions of higher education are already considering "bring your own device" strategies. Providing e-learning services via platforms like Moodle is all but imperative for a typical school or university. Even primary-school kids receive media education and are using powerful and appealing learning apps to practice basic competencies. (Math-training apps in particular are very popular.) This situation advantages learning analytics because the amount of digital data about learning-related activities and achievements is increasing steadily. However, the various sources are highly heterogeneous and widely disconnected, so there is no coherent and comprehensive basis for learning analytics on a global level that

© The Author(s) 2020

R. Jaakonmäki et al., *Learning Analytics Cookbook*, SpringerBriefs in Business Process Management, https://doi.org/10.1007/978-3-030-43377-2_5

allows educators to get new insights into the learning processes that occur in their own classrooms. To overcome this shortcoming, all available data must be brought together, aggregated, and linked to a suitable model of individual learning.

The European Lea's Box project addresses such problems by offering a central learning analytics platform that allows the various data sources to be linked and grants teachers a number of valuable analyses and reports across all sources. In addition, teachers receive easy-to-use notebook functions with which to enter achievements, activities, or learning progress manually. With only a bit of effort to set up a learning analytics solution, Lea's Box offers the most comprehensive approach and the deepest insights into the individual learning processes and learning states.

Preparation Time: Preparation time depends on the size of your organization, the number of courses, and the number of data sources. A simple scenario with a single teacher, a group of 20 students, and a single course can be set up in about an hour. Setting up a scenario for an entire school may take several hours.

5.2   Ingredients

- Account at Lea's Box platform, which is freely available
- Digital data sources (learning management systems, learning apps, assessment tools, manual records)
- A device with which to access the platform (computer, tablet)
- Network infrastructure, Internet access (not blocked by a firewall)

5.3   Preparation

Request an administrator account for your school or organization by clicking Lea's Box "View Product" button at http://www.learning-analytics-toolbox.org/tools/. This leads you to the Lea's Box website, where all contact information and procedures can be found.

Prepare a list of the teachers in your organization who should use the system, a list of students who should use the system, your courses, and a list of competencies addressed in each course. If you have these data in an Excel spreadsheet or csv file, you can automatically import the data into the Lea's Box system. For details and format specifications, refer to the manuals and guidelines on the Lea's Box portal.

Prepare your digital data sources so these sources can send their data to Lea's Box's central data store. This step may be tricky and require the help of your organization's technical support staff. With your administrator account, you will receive a URI and ID numbers for your data sources. These data sources export their data to the URI using a secret token as an identifier. This connection is established either through the simple web service API or the so-called xAPI, which is the current standard for the exchange of educational data.

5.4 Cooking

1. Set up your teachers, students, classes, courses, competencies, and external data sources. If this information is available in an electronic form, Lea's Box offers a conversion format to import these data. Alternatively, data can be added and edited directly through a web-based configuration tool that offers advanced functionalities in generating competence spaces and structures. With the administrator account, you can create as many users (teachers) as you want, and you can edit individual access rights.
2. Register your external data sources with the system using the configuration tool.
3. Carry out your lessons as usual, perhaps with an increased use of digital tools and devices. Send the data to the system.

5.5 Serving

The Lea's Box platform offers a variety of analytics and visualization features. For example, a teacher can view the aggregated data as "lattice graphs" (Fig. 5.1), which show students with similar achievements, learning levels, or interests. For example, the magnified area in Fig. 5.1 shows that Anna, Maria, Phil, and Sebastian have competencies C07, C08, C09, C15, and C16. Andreas is a bit further and also holds

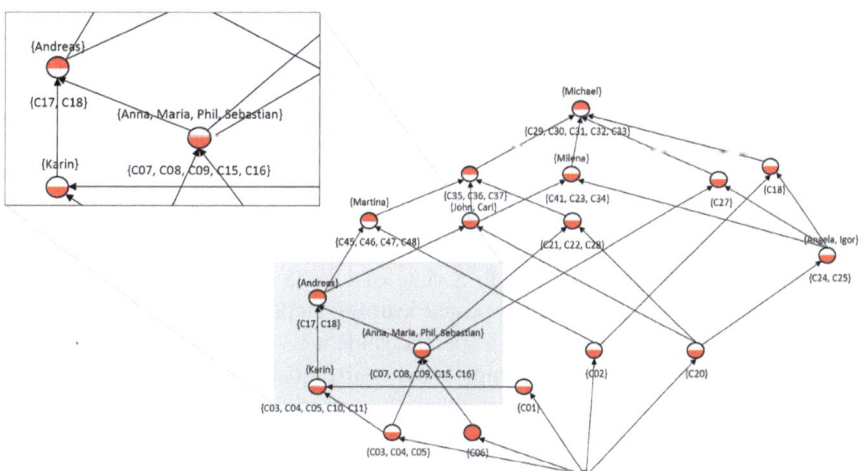

Fig. 5.1 Students' competencies presented in a lattice graph

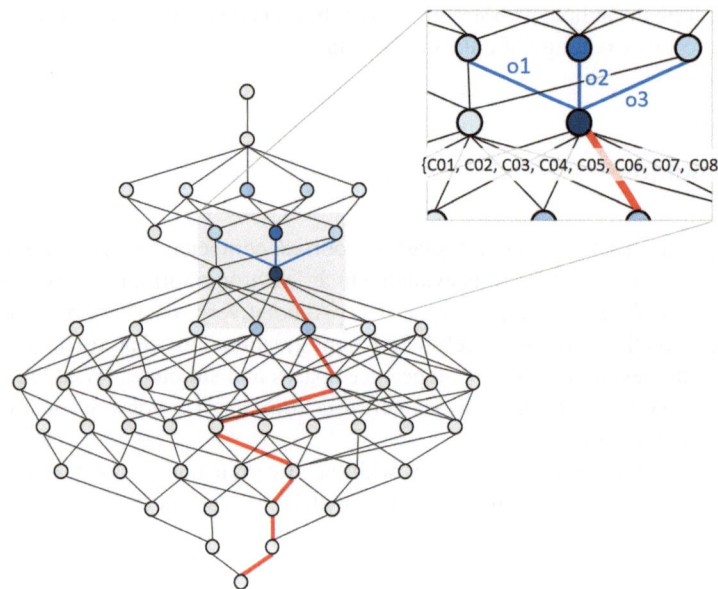

Fig. 5.2 Example of a student's competence space

competencies C17 and C18. Thus, you can identify subordinate and superordinate groups, relationships, and learning paths.

A similar visualization is "Competence Spaces" (Fig. 5.2), which shows the nonlinear structure of a course's competencies. The bottom of the diagram depicts a course's (or learning domain's) starting point, where the learners hold none of the competencies. The colors indicate the probability that a student has the competencies, where a darker color indicates higher probability. The lines show possible learning steps from one state to another—that is, the possible learning paths through the course. The example in Fig. 5.2 shows the concrete learning paths of a specific students (thicker red line) and the learning state with the highest probability (dark blue). The magnified area shows the competencies (C01–C08) this student has, as well as the optimal next learning steps (o1–o3).

In Fig. 5.2, the easiest competencies (or skills) are at the bottom and the most difficult are on top. The lines indicate possible learning paths to take from a state of having none of the competencies to a state of having all of them. This type of visualization also allows the next optimal learning steps for each individual to be identified, and it forecasts whether a specific student is at risk of failing the course because he or she is unlikely to complete the course goals in the remaining time.

Another visualization is the so-called "Learning Landscape," as shown in Fig. 5.3. The screen shows the students, their assignments, and their achievements in a landscape-like way. For example, Fig. 5.3's left panel shows a number of the course's exercises clustered based on their difficulty (or, in other words, the

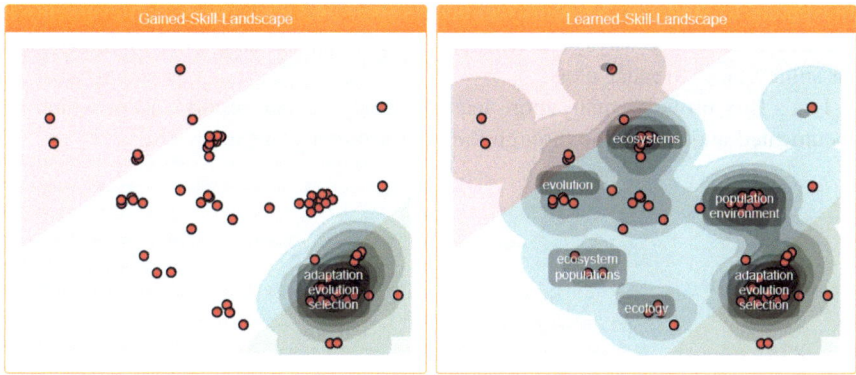

Fig. 5.3 Students' "Learning Landscape"

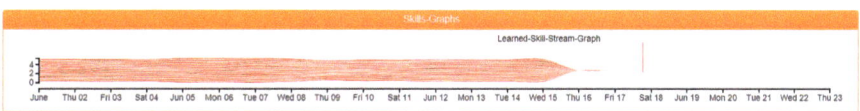

Fig. 5.4 Course progress timeline

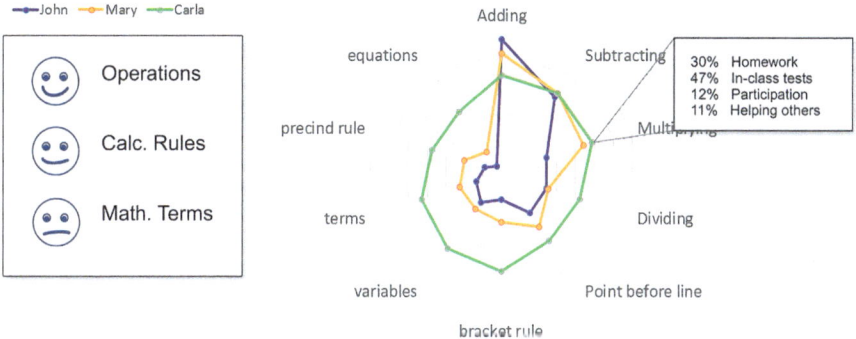

Fig. 5.5 Example of an open learner modelling visualization

probability that an exercise was mastered correctly) shown in a mountain-like height-scale—the darker the area, the higher the probability (or lower the difficulty level). The left panel shows the underlying competencies and the degree to which the students have them. With an interactive timeline (Fig. 5.4, lower panel), teachers can trace the details of the course's progress, including the individual students, the exercises they mastered over time, and the competencies they acquired.

A distinct feature of the Lea's Box platform is the Open Learner Modelling feature (OLM, Fig. 5.5), a rich platform with which to make analyses and results transparent to students. Students can access their and their peers' achievements in intuitive and suitable ways. The teacher can select the suitable types of charts, from bar smiley faces to radar plots.

Lea's Box provides many more features, analyses, and reports, all of which are documented and accessible through the platform that you can try out.

Chapter 6
Explore your Online Course: Visualization of Resource Use and the Learning Process

Abstract Suppose you developed a blended learning concept, also known as "flipped classroom", where your students can access content and continue learning at home, in addition to learning in a classroom. Now you just finished this course for the first time using the Moodle platform, and you would like to revise how it went and what aspects could still be improved. You are especially interested in when the students learned, how often they learned, whether they used all the provided resources, and how well they performed. This recipe offers an example how a teacher can gain insights into students' behavior in an online platform during the course. Using a tool for visualizing learning activities assists teachers to get a better overview of typical patterns in students' learning and performance. This can help to further improve the course content and your teaching methods.

Keywords LEMO2 CourseExplorer · Moodle course analytics · Visualization

6.1 Appetizer

Suppose you developed a blended learning concept (e.g., an inverted classroom) for your course and used the Moodle platform to manage the course online in addition to your regular offline/face-to-face class. Now you would like to see how the overall course went so you can improve the course design. Did students adapt well to the idea of a blended classroom? Did they use all the extra learning material you provided? When did they learn the most? Did they prepare for the face-to-face classes, or did they postpone active learning until the final examination? Were some resources heavily used and some that were rarely used? Did students use the quizzes you provided each week to check their knowledge level? Did students' performance improve over the course period? Did students use the Moodle online forum for collaboration or just as another source of information? These and more questions may arise based on the design of the course.

The learning analytics dashboard "LEMO2 CourseExplorer" provides multiple options for answering your questions and gives you an overview of how students

© The Author(s) 2020

R. Jaakonmäki et al., *Learning Analytics Cookbook*, SpringerBriefs in Business Process Management, https://doi.org/10.1007/978-3-030-43377-2_6

used all of the resources. LEMO2 provides various intuitive views on learning activities over time and visualizes students' performance and typical learning steps.

In addition to seeing an overall trend, you might also use filters to focus on your specific areas of interest, such as whether the time allowed for activities could be reduced whether the use of learning resources could be restricted to specific resources, or whether the learning behavior of students in different performance groups could be compared. The design of the LEMO2 CourseExplorer tool follows Shneiderman's (1996) well-known mantra for visualization and visual analytics: "Overview first, zoom and filter, then details-on-demand." While digging into your course, you might even find answers to questions you had not thought of that can empower you to re/design your course to do a better job of reflecting your students' needs.

Preparation Time: Several hours

6.2   Ingredients

For this appealing recipe, you need the following ingredients:

- Your Moodle course
- Your students' consent to use their data[1]
- The LEMO2 CourseExplorer tool

6.3   Preparation

6.3.1   Installation of the LEMO2 CourseExplorer

By clicking "View Product" at http://www.learning-analytics-toolbox.org/tools/—LEMO2, you can obtain the installation files and a user's and installation guide that leads you through the installation of the CourseExplorer.

6.3.2   Export Your Moodle Course

You can add any Moodle course, whether still active or already finished, to LEMO2 CourseExplorer. If the Moodle course was developed primarily to distribute lecture notes and supplementary materials, you can see how various learning materials are being used. However, with a richer course (e.g., organizing learning resources in

---

[1]Please check your compliance with your university's data privacy regulations.

chapters, having a variety of learning resources like web links for further reading, ordering resources chronologically, adding forums and quizzes), you can attain more insight into the learning processes that take place.

To access the activity data in your Moodle course, you need a Moodle backup file, which must be approved by the Moodle system administrator. At this point, you will probably have to discuss about compliance with your university's data privacy regulations experts. The data used in LEMO2 CourseExplorer is anonymous (i.e., LEMO2 does not store any data that could be related to individual students), so it should be possible to use LEMO2 CourseExplorer at most universities.

Finally, you need your students' consent to analyze their learning activities with the CourseExplorer. We suggest that you show students the benefits of learning analytics by presenting your goals and rationales for wanting to reflect on the learning processes that will be visible from their activity data.

6.3.3 Importing Your Data into the CourseExplorer

The final step of your preparation is to import your course data into the LEMO2 CourseExplorer. We provide a short explanation of how the LEMO2 CourseExplorer works, but for the most current version, refer to the user's guide.

To import a course from a Moodle backup, select "Import Course Data" from the top menu, upload the backup from your file system (Fig. 6.1), and see the status of your import.

After your backup was imported successfully (Fig. 6.2), you can select an analysis from the top menu and start exploring your course.

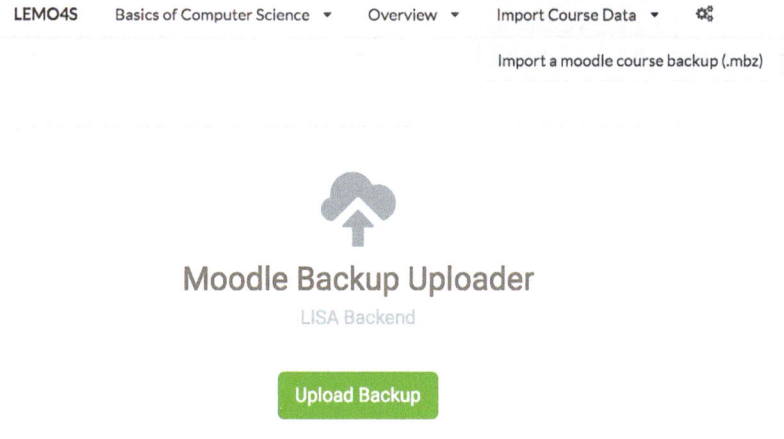

Fig. 6.1 Import Moodle course data screen

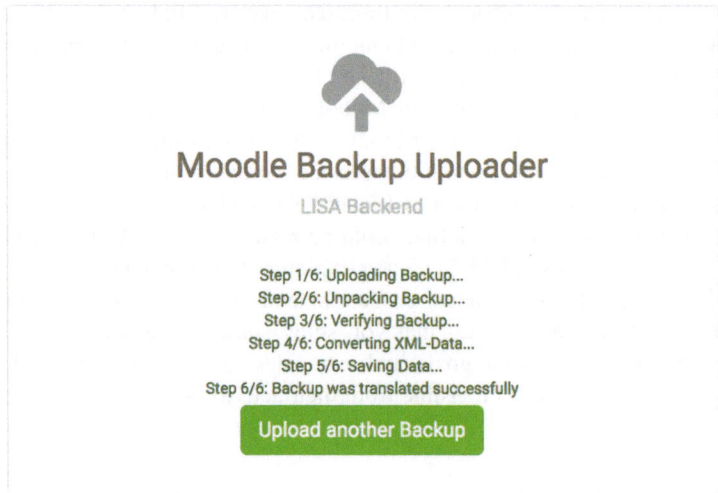

Fig. 6.2 Import status screen

6.4 Cooking

6.4.1 Essential Kitchen Tool: The Dashboard

We provided a brief overview of the LEMO2 CourseExplorer in the preparation section. Now let us get more familiar with the CourseExplorer.

In the Top Menu are several options to configure the CourseExplorer to your course (Fig. 6.3). Start by choosing your course from the drop-down menu. Then select your basic configurations for the analysis. If you are uncertain about which configurations to choose, we recommend that you explore your students' activities over time and compare the activities to learning objects to get a nice overview of your course.

A graph section shown at the center of the LEMO2 course explorer is where your results are visualized course (Fig. 6.3). The graph section will give you feedback on whatever nuances you add to your analysis. You can use the cursor (the gray window below the graph in Fig. 6.3) to interact with the graphs and get the full experience of your analysis.

In Fig. 6.3, below the graph section is a detail section with a list of detailed information about the analysis results. You can sort it by clicking a header, such as name, type, users, activities, average activities per user, and peak activity in terms of time.

On the right-hand side is an additional filter section that offers four filters for analyses (Fig. 6.3). Using the filter section, you can set a data range, choose certain students by grade, and select specific learning objects or types of learning objects.

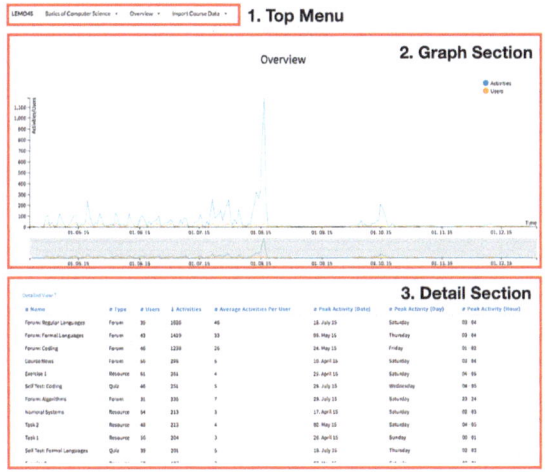
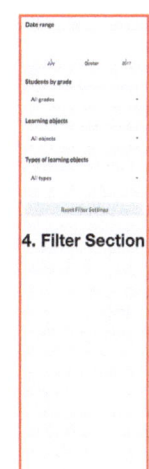

Fig. 6.3 LEMO2 CourseExplorer sections

6.4.2   Different Flavors of Course Analysis

Just as a proper dish has many flavors, LEMO2 CourseExplorer incorporates various analyses. Our analyses include an overview of students' learning over time, students' use of learning objects, use patterns in a calendar heat map, daily activity in a box plot, navigating between learning objects in an activity flow, each student's test performance, and learning achievement as a distribution. If you are already accustomed to these analyses, you can dig into a specific analysis. However, if you would like to know more about what each analysis entails, you can either keep reading this section, which contains descriptions of each analysis, or refer to the user guide.

Overview

The line graph in Fig. 6.4 shows both the number of activities students have completed and the number of students. The slider below the graph section allows you to zoom into a specific timeframe. Zooming can also be done by scrolling your mouse in the gray window. Hovering your cursor over the line graph reveals more detailed information about the interactions. Below the graphs is a list of details about the learning objects, such as name, type, number of users, and number of interactions.

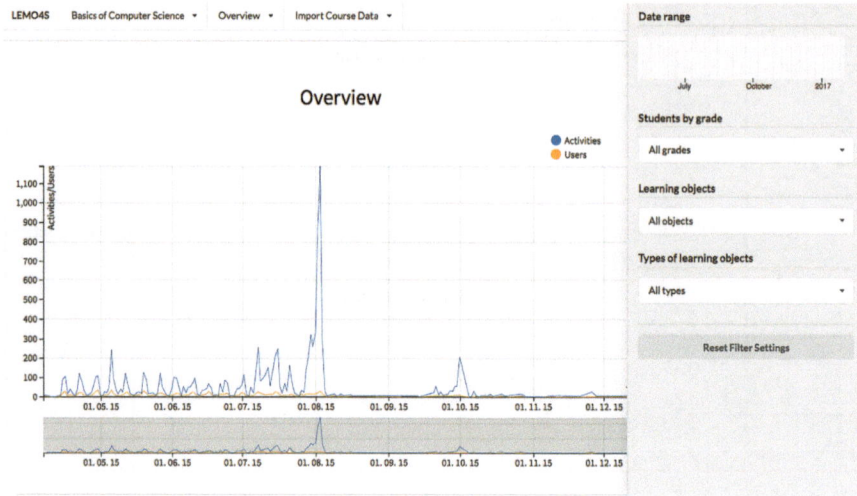

Fig. 6.4 Overview of activities and users on a timeline

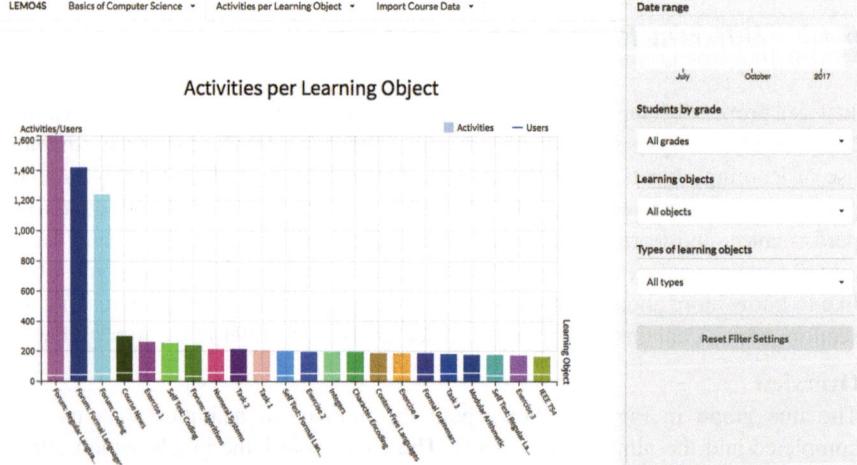

Fig. 6.5 Activities compared to learning objects

Activity per Learning Object

Learning objects are also used in combination with activities. As Fig. 6.5 shows, a line graph and bar chart are combined to show the number of users and interactions for each learning object. Each learning object has its own color, regardless of the kind of analysis you are looking at, which will help you to identify it in various types of visualizations. Below the graphs is a list of details about the learning objects, such as name, type, number of users, average interactions per user, and average usage time.

Calendar Heatmap
A heatmap is used to blend a calendar with the number of activities performed (Fig. 6.6). The number of activities per day is displayed such that darker shades represent more activities performed. The calendar shows the course's activities for the full year.

Activity Over Weekdays (Boxplot)
The calendar view in Fig. 6.6 can present the whole year or a single day. The activity over weekdays in a boxplot provides detailed information about students' activities. Figure 6.7 displays each weekday as a box plot containing information on the number of activities on that day. Unlike the calendar view, the boxplot shows the distribution of activities on specific weekdays.

Activity Flow
The activity flow analysis shows the coherences and associations between certain learning objects as a chord diagram (Fig. 6.8). The colorful diagram also shows where most students navigate within the learning materials. The top left of the chord diagram has a legend with incoming, outgoing, and both directions. Green incoming connections and nodes indicate that the selected object was visited from a green node (i.e., shows what was the previous activity of the student), whereas students navigate

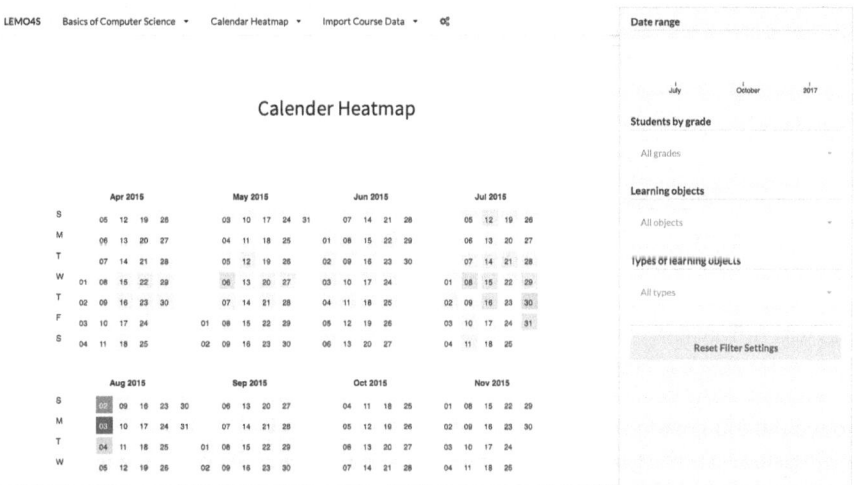

Fig. 6.6 Activity Calendar

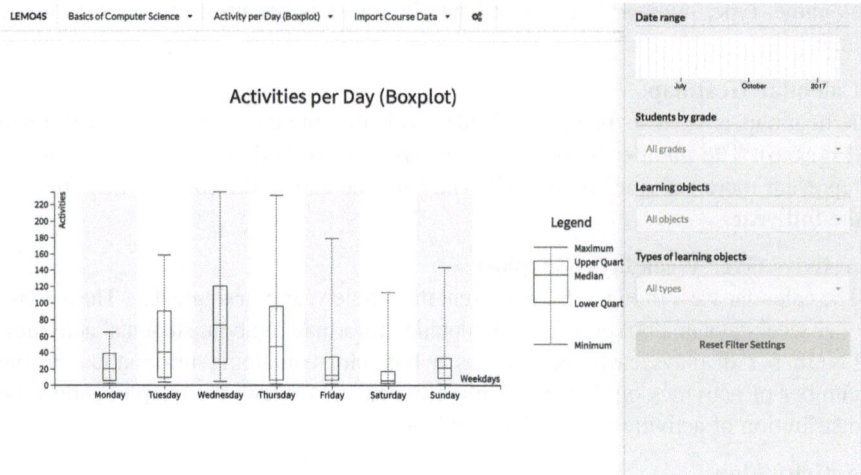

Fig. 6.7 Activity per day (boxplot)

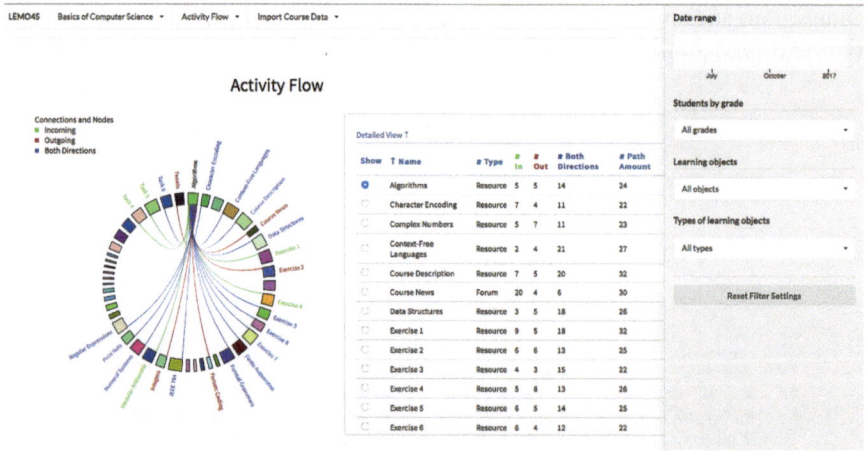

Fig. 6.8 Activity flow

to red outgoing nodes from the selected spot (i.e., shows what was the next activity of the student). Blue nodes and connections indicate that learners navigate to and from the selected object. A detailed view is shown on the right side so you can select a specific learning object using a radio button.

Test Performance per Student
The CourseExplorer provides a visual analysis so you can compare your students' overall performance. As in Fig. 6.9, each student's performance in all quizzes and

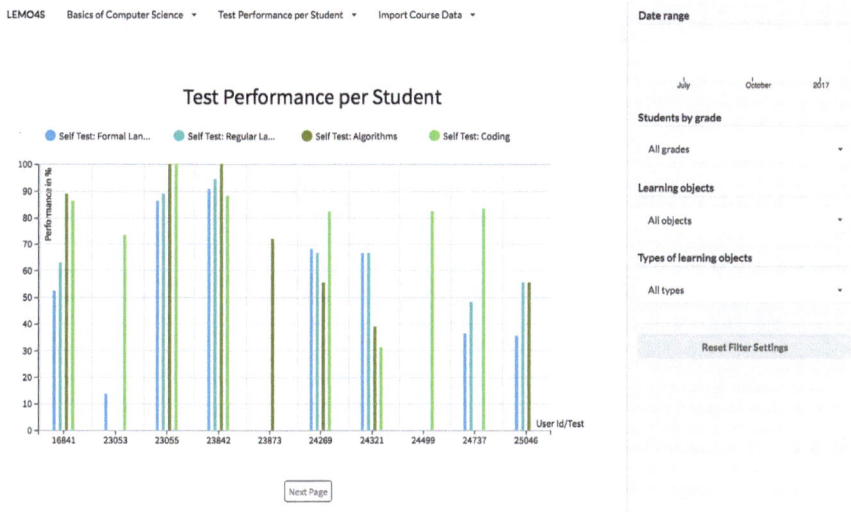

Fig. 6.9 Test performance per student

tests is displayed as percentage of full points (y-axis). To avoid data privacy issues, students' names are anonymized with a random number (x-axis). You can also disable or enable a specific quiz or test by using the color-coded tabs on the top of the graph.

Performance Distribution per Test
Each student's individual data can give you an idea of which tasks were more difficult, and using the "Quizzes" view (Fig. 6.10) shows the students' performance on the tests or quizzes in the course. The number of students enrolled in the course is indicated in the top left corner of the graph.

The bar graph is separated into five sections:

1. students who have participated in the selected test and achieved a score between 1% and 59%
2. students who achieved a score between 60% and 69% (letter grade D),
3. students who achieved a score between 70% and 79% (letter grade C),
4. students who achieved a score between 80% and 89% (letter grade B) and
5. students who achieved a score greater than 89% (letter grade A).

When you hover your mouse cursor over the bar graph, the number of students who have attempted the test is shown on the top of the bar so you can review detailed information on each group of students.

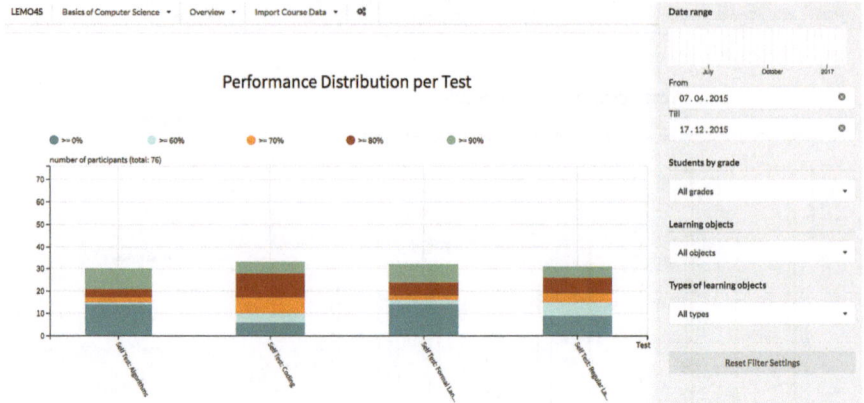

Fig. 6.10 Performance distribution per test

6.4.3   Seasoning

While basic analyses are great for getting an overview of your course, filters allow you to do some fine tuning to get more sophisticated results. You can find the filters on the right-hand side of the dashboard. The graph will change based on your selections in the filter.

The date-range filter allows you to choose which time period to focus on by using the slider tool in the filter section. To give you some information while you explore the data, the slider tool contains a preview of your graph. Another way to refine the data is to filter students by letter grade (A, B, C, D, and F). Your analysis can also be refined to regard only certain learning objects.

The drop-down menu contains all of the learning objects in your course. If you want to explore a specific type of learning object, you can use the "Types of learning objects" filter.

6.5   Serving

In the presentation of the dish, what counts is not just the taste but also the visual appearance. Below you can find three themes for presenting your dish. We'll use an example course to navigate through these themes, stated in the form of questions.

6.5.1 How Do Students Use Their Learning Resources?

Course designers often include a lot of ingredients (e.g., text files, online videos) in the study material. Here is where you can see whether these ingredients are presented properly and determine what might spice up the course in the future. To do this, we'll delve into the resources of LEMO2 CourseExplorer.

The first big question is whether our guests (i.e., students) have been enjoying our course. To answer this question, you need to know whether all the ingredients have been consumed and enjoyed by determining how they have been used, so we ask three sub-questions:

- Are the students using the learning resources regularly?
- What resources are used more frequently than others?
- What resources are seldom used?

Are the Students Using the Learning Resource Regularly?
LEMO2 CourseExplorer provides a visual representation of how learning resources are used. For instance, look at the fresh ingredients (e.g., recent online video links) that were used in your meal: have they been rated as a harmonic addition to the whole course or not? So let's see whether our students are using their learning resources regularly. When you log in, you will see "Overview," which shows the learners' activity over time, as shown in Fig. 6.11.

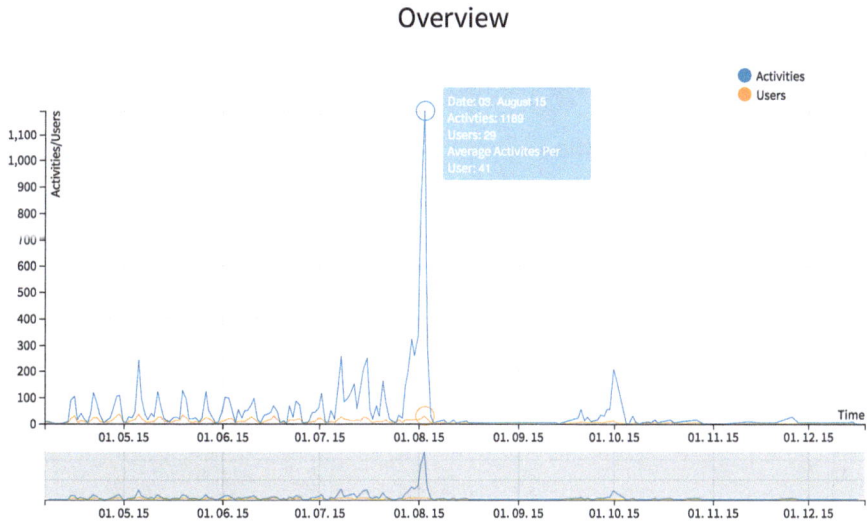

Fig. 6.11 Number of activities and users over time

The graph in Fig. 6.11 uses two line graphs to show the students' overall activities between April and December. One graph shows the number of users, while the other graph shows the number of activities in the course. When you hover your cursor over the graph, the peak's details are shown. For example, on August 3, the peak of the activities line, 29 of the course's students were online, and there were 1189 activities, thus indicating that, even though only 29 users were active on this date, they were quite active, with each student accessing 41 learning objects on this date.

This view does not show us which learning objects were accessed, but we know that the highest number of activities was recorded on August 3 and that course was accessed regularly from April to the beginning of August.

The area of interest can be zoomed in on as well, as shown in Fig. 6.12.

The activities during this time vary, but the number of users who were accessing the course appears stable. We can also see that no activities are shown between August and the end of September but that some activities reappeared in October (Fig. 6.13).

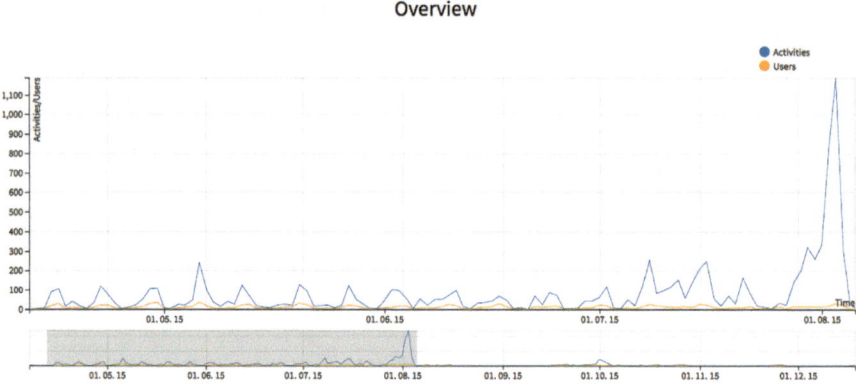

Fig. 6.12   Zoom-in view of activities

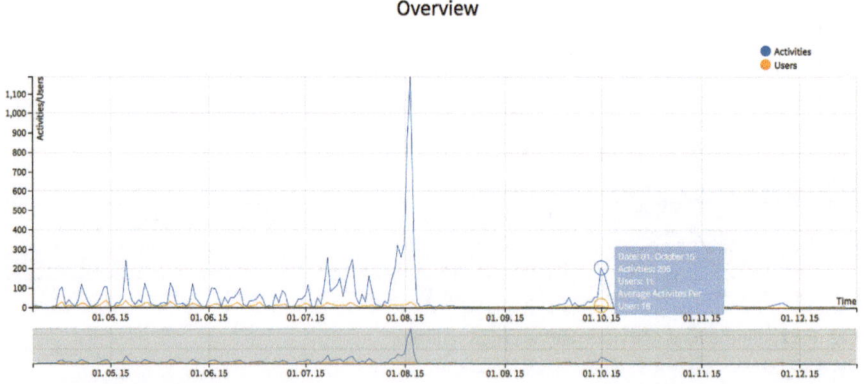

Fig. 6.13   Activities in October

Now we can trace back and think about what might have caused this result, that is, why there is another activity shown after about a month of dormancy.

The high points in the number of activities could be the result of having two final examination periods (one in July or August and the other in September or October), as this course offers two opportunities for students to take their final exam. Most of the students seem to have taken the exam in the first period, before students in the second exam period accessed the online Moodle course data.

What Resources Are Used More Frequently Than Others?
To see which learning object has the highest number of visit, you can look at the detailed view under the "Overview" visualization. When sorting the activities in descending order, we can see that *Forum: Regular Languages* was accessed the most, with 35 users (Fig. 6.14).

The number of times this forum was accessed was 1626, indicates that the students accessed it an average of about 46 times during the course. The peak activity was on July 18 between 3:00 and 4:00, and this learning object was frequently visited on Saturdays.

Two other forums that were administered in this course (*Forum: Formal Languages* and *Forum: Coding*) also showed large numbers of accesses by students. (Note that this does not indicate that these top three learning objects were accessed by the most users, just the most times.) Let's look at the detail view sorted by the number of users shown in Fig. 6.15.

In Fig. 6.15, the same data has been sorted by users. Here we can see that *Exercise 1*, *Course News*, *Task 1*, *Modular Arithmetic*, and *Numeral Systems* are the top five learning objects, as each was accessed by more than 54 students somewhere between three and five times on average.

These five learning objects were accessed in April, specifically around weekends. As the course description is presented at the beginning of the semester, it not

Detailed View ↑

# Name	# Type	# Users	↓ Activities	# Average Activities Per User	# Peak Activity (Date)	# Peak Activity (Day)	# Peak Activity (Hour)
Forum: Regular Languages	Forum	35	1626 ✔	46	18. July 15	Saturday	03 - 04
Forum: Formal Languages	Forum	43	1419 ✔	33	09. May 15	Thursday	03 - 04
Forum: Coding	Forum	46	1238 ✔	26	24. May 15	Friday	01 - 02
Course News	Forum	56	298	5	10. April 15	Saturday	03 - 04
Exercise 1	Resource	61	261	4	25. April 15	Saturday	04 - 05
Self Test: Coding	Quiz	46	251	5	29. July 15	Wednesday	04 - 05
Forum: Algorithms	Forum	31	235	7	29. July 15	Saturday	23 - 24
Numeral Systems	Resource	54	213	3	17. April 15	Saturday	02 - 03
Task 2	Resource	48	213	4	02. May 15	Saturday	04 - 05
Task 1	Resource	56	204	3	26. April 15	Sunday	00 - 01
Self Test: Formal Languages	Quiz	39	201	5	18. July 15	Thursday	02 - 03
Exercise 2	Resource	50	197	3	02. May 15	Saturday	00 - 01
Integers	Resource	51	196	3	17. April 15	Friday	02 - 03
Character Encoding	Resource	51	193	3	17. April 15	Saturday	00 - 01

Fig. 6.14 Resources with high levels of activity

Detailed View ↑

# Name	# Type	↓ Users	# Activities	# Average Activities Per User	# Peak Activity (Date)	# Peak Activity (Day)	# Peak Activity (Hour)
Exercise 1	Resource	61	261	4	25. April 15	Saturday	04 - 05
Course News	Forum	56	298	5	10. April 15	Saturday	03 - 04
Task 1	Resource	56	204	3	26. April 15	Sunday	00 - 01
Modular Arithmetic	Resource	54	178	3	17. April 15	Friday	01 - 02
Numeral Systems	Resource	54	213	3	17. April 15	Saturday	02 - 03
Character Encoding	Resource	51	193	3	17. April 15	Saturday	00 - 01
Integers	Resource	51	196	3	17. April 15	Friday	02 - 03
Exercise 2	Resource	50	197	3	02. May 15	Saturday	00 - 01
Submission 1	Assign	50	50	1	07. May 15	Wednesday	13 - 14
Context-Free Languages	Resource	49	187	3	09. May 15	Friday	01 - 02
Task 3	Resource	49	183	3	30. May 15	Friday	03 - 04
Exercise 3	Resource	48	170	3	30. May 15	Friday	01 - 02
IEEE 754	Resource	48	163	3	11. May 15	Saturday	04 - 05
Task 2	Resource	48	213	4	02. May 15	Saturday	04 - 05
Formal Grammars	Resource	47	184	3	14. May 15	Sunday	04 - 05
Exercise 4	Resource	46	184	4	05. June 15	Friday	01 - 02
Forum: Coding	Forum	46	1238	26	24. May 15	Friday	01 - 02

Fig. 6.15 Learning objects with high numbers of users

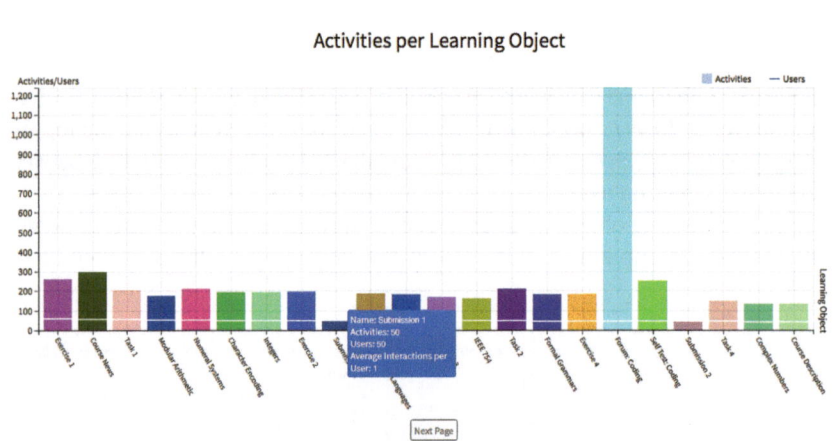

Fig. 6.16 Number of activities per learning object, sorted by number of users

surprising that *Course News* was accessed at the beginning of the semester. The first face-to-face class was first held on April 15, and we see that students accessed the course description a little less than a week before that, on April 10. The learning objects with large numbers of users have their peak activity dates around April 10 to April 25, on Fridays, Saturdays, and Sundays, while the peak hours were from midnight to 5:00 a.m.

Let's try a different visualization to look more deeply into the learning objects' activities. Figures 6.16 and 6.17 show all of the learning objects in this course. As it is still sorted by users, Fig. 6.16 shows that *Exercise 1* has a lead in terms of the

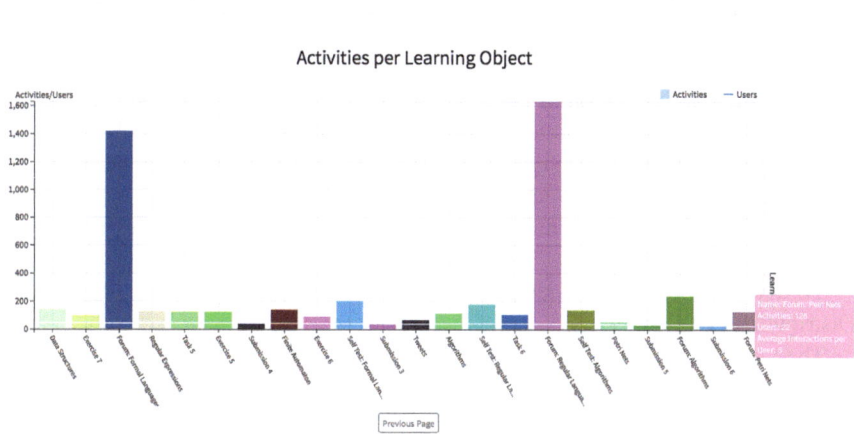

Fig. 6.17 Number of activities per learning object, sorted by number of users, and zoomed in *Forum: Petri Nets*

Detailview ↑

↕ Name	↑ Type	↕ Users	↕ Activities	↕ Average Interactions Per User	↕ Peak Activity (Date)	↕ Peak Activity (Day)	↕ Peak Activity (Hour)
Submission 1	Assign	50	50	1	07. May 15	Wednesday	13 - 14
Submission 2	Assign	46	46	1	27. May 15	Thursday	12 - 13
Submission 3	Assign	39	39	1	21. May 15	Monday	05 - 06
Submission 4	Assign	41	41	1	11. June 15	Monday	05 - 06
Submission 5	Assign	34	34	1	18. June 15	Sunday	18 - 19
Submission 6	Assign	26	26	1	16. August 15	Thursday	17 - 18

Fig. 6.18 Assignment learning objects

number of users. The previous visualization showed that three forums have significantly high number of accesses. However, *Forum: Petri Nets* ranked at the bottom, with only 22 users (Fig. 6.17).

Another interesting aspect of this visualization is that *Submission 1* was ranked comparatively high, with 50 users (Fig. 6.18), but users accessed the other submission learning objects only once. This easy to explain, as the course is designed so each student can submit only one assignment. According to the syllabus, students are eligible to take a final exam only when they have completed Submissions 1 through 4. The detail view of the visualization shown in Fig. 6.18 indicates that 39 students completed Submissions 1–4, so a maximum of 39 students were able to take the exam to pass this course.

The syllabus also indicates extra credit for students who completed Submissions 5 and 6. Figure 6.18 indicates that 26 students completed all assignments and are eligible for extra credit.

What Resources Are Seldom Used?

Similar to how we answered the question concerning which resources are used most frequently, we can find which resources were seldom used using the "Overview" and "Activities per Learning Objects" visualizations. Before looking into the same bar graph to get an idea of overall class size, let us visit another visualization for a quick view.

To find the number of students enrolled in this course, we jump to "Performance Distribution per Test."

The top-left corner of Fig. 6.19, right below the legend of distribution, shows that there were 76 students enrolled in this course. As our focus now is the resource that was not used much, we go back to the previous visualization ("Activities per Learning Object") and find that *Forum: Petri Nets* and *Submission 6* are ranked at the bottom, with fewer than 40% of the enrolled students having accessed these two learning objects (Fig. 6.20).

Figure 6.20 also shows that these two seldom-used learning objects were accessed late in the semester, as was the case for other learning objects with few users. As we have observed that a large number of students accessed the course data at the beginning of the semester and fewer near the end, it might be instructive to arrange the learning objects accordingly. With the peak activity hours on Saturday morning and on Thursday evening, the usage patterns of the resources that were accessed by the fewest students also differ from those of the resources that were accessed most frequently.

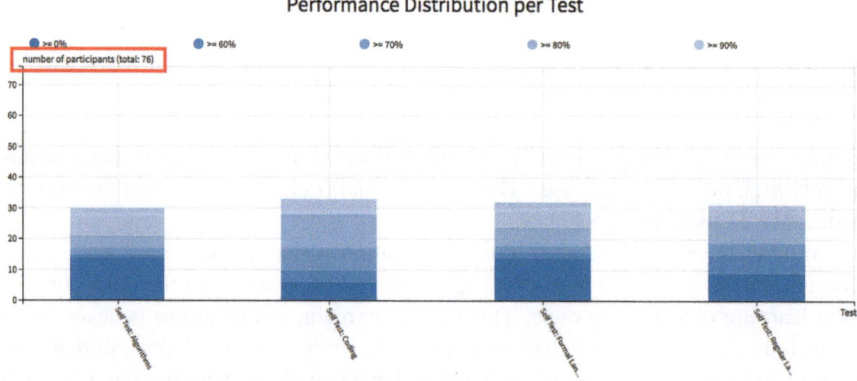

Fig. 6.19 Performance distribution for course participants

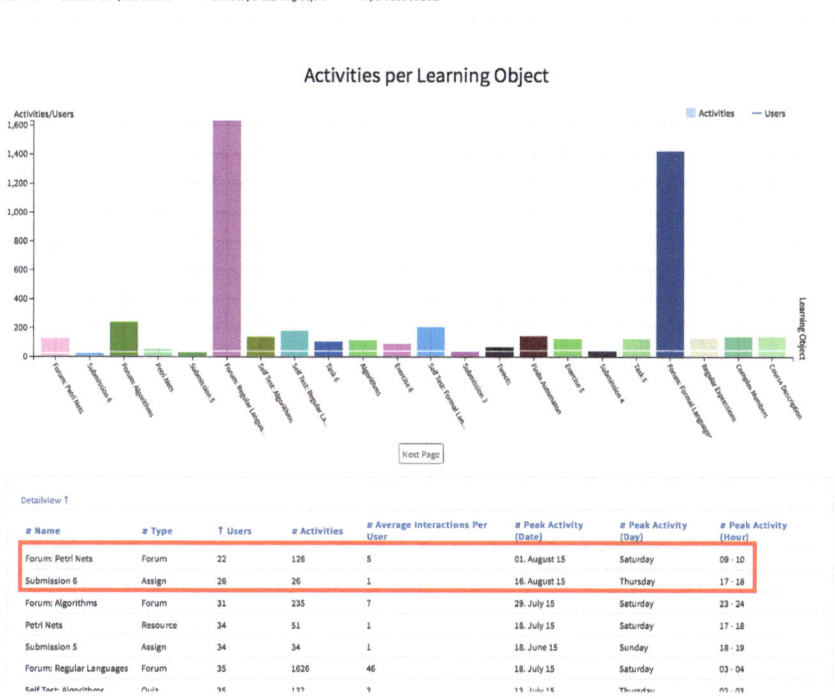

Fig. 6.20 Activity displayed per learning object

As a course designer and teacher, you may want to end this section by thinking about how you would answer three questions:

- Are resources that students must use to learn key concepts offered in the prime time of students' attention?
- How can one motivate students to continue rather than dropping out in the middle of the course?
- Are there other resources that can be added to help students learn?

6.5.2 Are There Patterns to Describe Students' Online Learning?

The type and quality of the study material is the ultimate decision-maker, yet seasonal variables may affect when students try out the various study materials. Following such a pattern may make our course more harmonic with students' own patterns and habits. Therefore, our second question deals with students' pattern of learning with respect to when and how they learn. We address two sub-questions to determine when our students prefer being active and adventurous:

- When do students learn?
- In what order do students navigate through learning resources?

When Do Students Learn?

We have observed that the beginning of the semester is the best time to offer our special menu items so we can ensure that students are equipped with the core knowledge and skills they need to continue learning. With only the active students in mind, this question investigates whether the date, time, and day affects the users. When a big event like an exam is around the corner, are our users be more prone to try our appetizers and desserts? When a face-to-face class is offered, is that the day that students access the Moodle course?

To see when students are active in using Moodle course data, we can use calendar view (heatmap). This visualization shows the number of activities that are centered on a specific month, week, or even day. Figure 6.21 shows that August 2 and 3 are the most active dates, and some frequent activities are observed in July.

The choice of these activities may be due to the first period of evaluation, and the high number of activities may correspond to preparation for the exam. Frequent

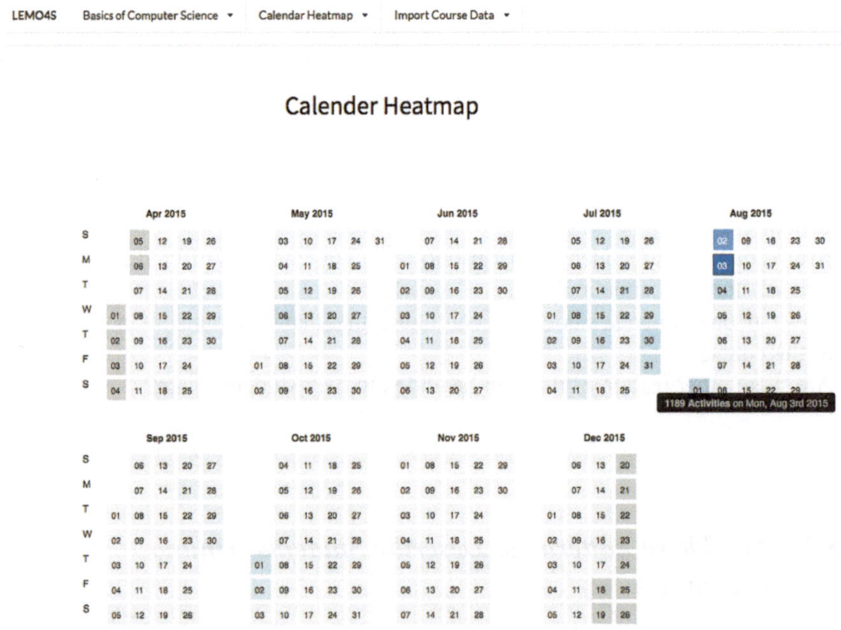

Fig. 6.21 Calendar heatmap for course activities

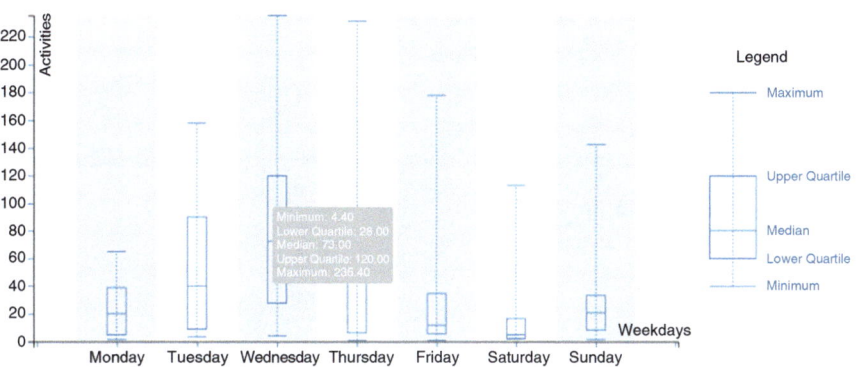

Fig. 6.22 Activities displayed in a boxplot

visits were also made on Wednesdays (May 6, July 8, July 15, and July 29), as shown in "Activities per Day" in Fig. 6.22.

Activities on Wednesdays had the highest median at 73 activities, a minimum of around four activities, and a maximum of 235 activities. As also shown in the calendar view, Wednesdays were the days with the highest number of activities overall, perhaps because the class was offered on Wednesdays, so students access the online course before the face-to-face class to prepare for it and after the class to review it.

In What Order Students Navigate Through the Learning Resources?

Related to the selection of study material, can we see some pattern in our students' choices? For instance, if a user took a short test, does he or she go on to do another exercise or take another test? Do users take the test after doing some exercises or do they access some reading materials and then take the test? These inquiries can be answered using the LEMO2 CourseExplorer "Activity Flow" visualization, shown in Fig. 6.23.

The visualization in Fig. 6.23 is sorted by the type of learning object, and *Submission 1* is selected. Students who take *Submission 1* tend to originate from *Course News*, *Task 2*, *Task 4*, *Task 5*, and *4 Forums*. After they take *Submission 1*, they go to *Submission 2* and *Submission 3*. They try out various resources before submitting the first assignment. As this course was designed to allow only those students who have completed all four submissions to take the exam, we can look for patterns of visits to the assignments.

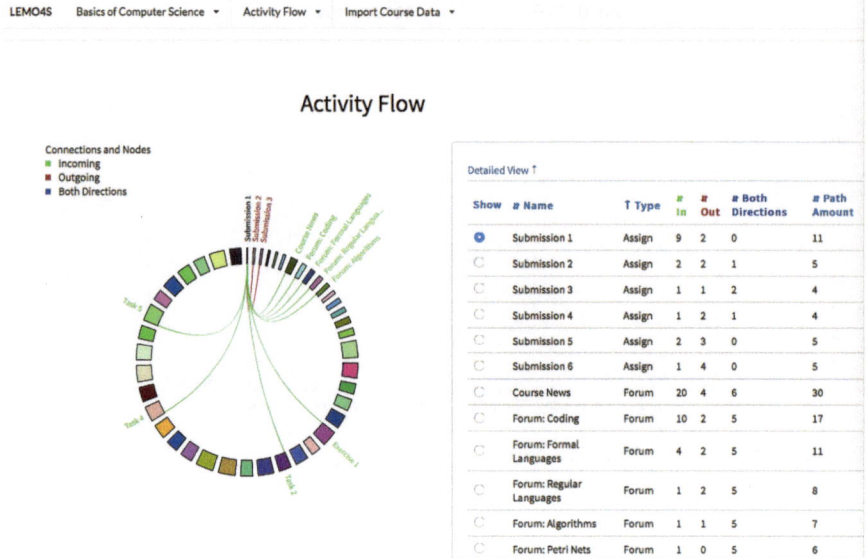

Fig. 6.23 Activity Flow for *Submission 1*

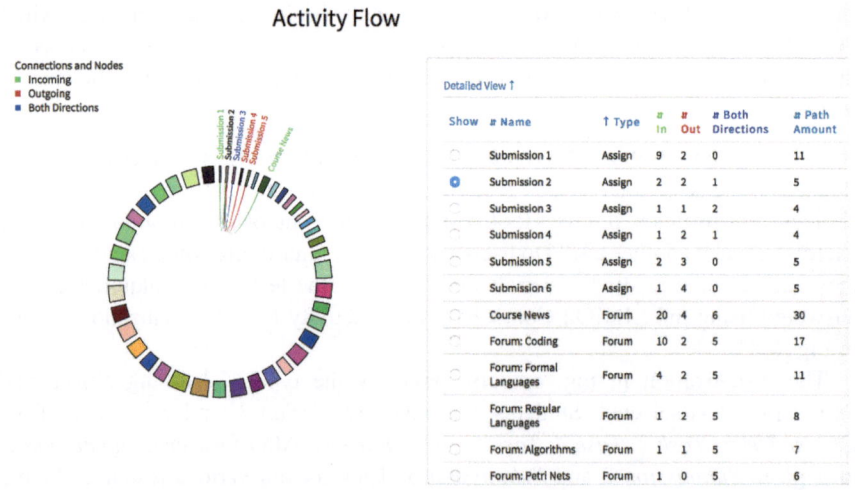

Fig. 6.24 Activity Flow for *Submission 2*

Submission 2 is reached from the *Submission 1* and from *Course News* (as we see in Fig. 6.24). Similar to the first assignment, students check *Course News* before taking the second assignment. After the second assignment, students travel toward assignments 3, 4 and 5. Unlike the pattern shown in the first assignment, Fig. 6.24

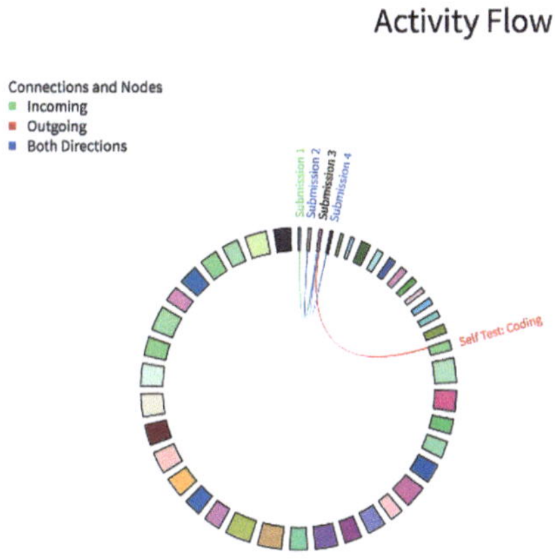

Activity Flow

Fig. 6.25 Activity flow for *Submission 3*

shows multiple assignments were submitted and that students travel to and from *Submission 3*. Unlike the first assignment, which was initiated prior to the second or the third assignment, there is a pattern of students coming back to the second assignment after the third assignment.

For assignment 3, we see a multidirectional pattern similar to that seen in assignment 2 (Fig. 6.25).

Figure 6.25 shows that students visited assignment 4 before and after assignment 3. From assignment 3, students also visited *Self Test: Coding*. Self tests are quiz items that students can use to test themselves. This visualization shows the relationship between assignment and quiz.

We see a similar pattern in Fig. 6.26 for assignment 4, which students visited after assignments 2 and 3. After completing assignment 4, students visited assignment 5 and *Self Test: Coding*, the latter to check their understanding of assignment 3, in addition to progressing to assignment 5.

The next section looks more deeply into how students perform in the Moodle course, but before leaving this section, you may want to ask yourself how you would answer three questions:

1. Since there are vast accesses before the exam and on the day of the offline course, what resources should be added or revised to improve the pattern of learning?
2. Is the pattern of learning objects visits corresponding to the course design?
3. What types of support should be given in the offline course to promote frequent visits to the online materials?

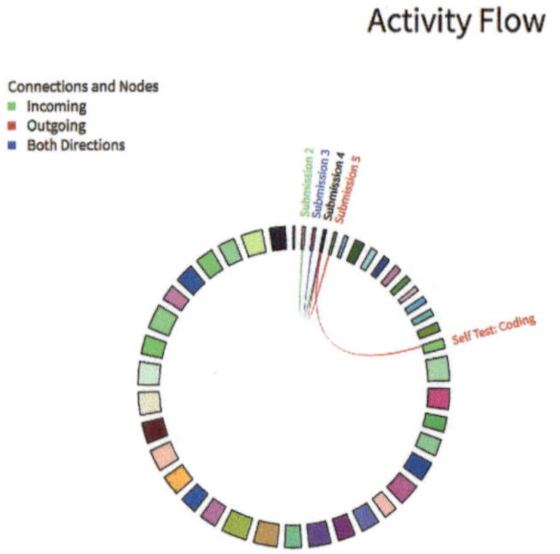

Activity Flow

Show	# Name	↑ Type	# In	# Out	# Both Directions	# Path Amount
	Submission 1	Assign	9	2	0	11
	Submission 2	Assign	2	2	1	5
	Submission 3	Assign	1	1	2	4
●	Submission 4	Assign	1	2	1	4
	Submission 5	Assign	2	3	0	5
	Submission 6	Assign	1	4	0	5
	Course News	Forum	20	4	6	30
	Forum: Coding	Forum	10	2	5	17
	Forum: Formal Languages	Forum	4	2	5	11
	Forum: Regular Languages	Forum	1	2	5	8
	Forum: Algorithms	Forum	1	1	5	7
	Forum: Petri Nets	Forum	1	0	5	6

Fig. 6.26 Activity Flow for *Submission 4*

6.5.3   How Do Students Perform?

The previous sections investigated patterns in the use of our resources and when students visited them. Based on this investigation, one can consider whether to redesigning some or all of the course's resources.

In this section, the students' actual performance is investigated. Taking tests and quizzes is a good way for students to check their knowledge level. This behavior tends to be shown by students with good learning-management skills and motivation to learn. By using a "Performance per Student" view, one can see how students are taking the course's tests and quizzes. For instance, four self-tests are administered in this example course (*Self Test: Coding*, *Self Test: Regular Languages*, *Self Test: Algorithms*, and *Self Test: Formal Languages*). The previous section shows some relations between two learning objects, assignment and self-test.

This section investigates two sub-questions regarding individual and overall class performance.

- Is any student struggling in taking a test?
- What is the overall performance distribution of the class's test scores?

Is Any Student Struggling in Taking a Test?
This question is designed to determine whether students are struggling with a quiz and, if so, if there is any support that teachers can provide by adding supplementary resources and forms of motivation.

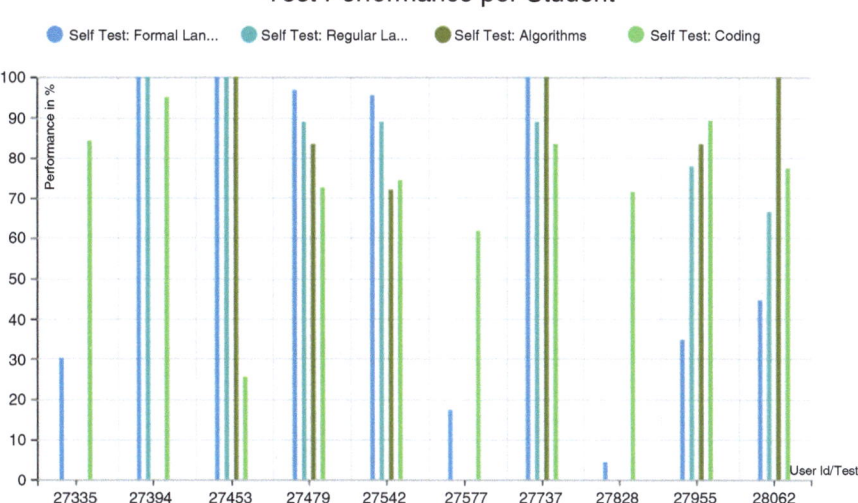

Fig. 6.27   Performance per student view

As Fig. 6.27 shows, students vary in terms of which tests they complete and in their performance on each test. Some students completed all tests with flying colors, while others took only one or two tests and had low scores.

By flipping through this view's pages for all students, we know that 4 students out of 11 did not take all tests and that. Out of that ten, three students took only two tests. We can also focus on details about each exam or student separately. For example, five students did not perform *Self Test: Formal Languages*, and scored below 50% on this test, and student 27,577 seems to overall struggle the most with the tests.

What Is the Overall Performance Distribution of the Class's Test Scores?

As teachers, we would like to motivate students who are not performing well and to mitigate any dropouts. The LEMO2 CourseExplorer provides a perspective of the class's overall performance in each quiz or test, as shown in Fig. 6.28.

Seventy-six students are enrolled in the course, but only between thirty and thirty-three students took the tests. Among four tests that are administered, *Self Test: Formal Languages* and *Self Test: Algorithms* had the highest failure rate, and *Self Test: Coding* had the lowest failure rate.

In the "Performance Distribution per Test" view, we can see that *Self Test: Coding* is well distributed, with around 15% of students scoring 90–100%, 33% scoring 80–89%, and 21% scoring 70–79%. *Self Test: Algorithms* and *Self Test: Formal Languages* must be considered difficult since students attempted them but could not complete them successfully.

Such investigations using LEMO2 CourseExplorer enables teachers to dig into students' progress so they can support their students.

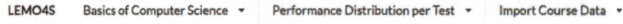

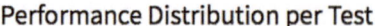

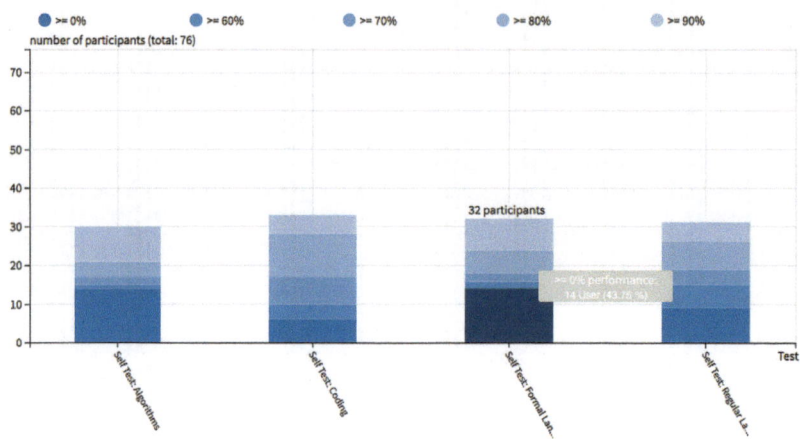

Fig. 6.28 Performance distributions per test

As a course designer and teacher, consider asking yourself four questions:

1. Is there any way to improve students' success in the self-administered quizzes and tests they take to check their progress?
2. What are the causes when most of students fail a particular quiz or test?
3. Are there any observable differences in the difficulty levels of quizzes and tests?
4. What supplementary resources should be added to help students improve their quiz and/or test results?

Reference

Shneiderman, B. (1996). The eyes have it: A task by data type taxonomy for information visualizations. In *Proceedings 1996 IEEE Symposium on Visual Languages* (pp. 336–343). Proceedings of IEEE, Boulder, CO.

Chapter 7
Understanding Students' Online Behavior While They Search on the Internet: Searching as Learning

Abstract Informal learning activities, like searching for information on the Internet, can enhance students' learning. However, not everything we find on the Internet is factual. If teachers could understand where the students look for information during their course, they could help them improve the quality of their informal learning through the Internet. This recipe shows how to gain insights into students' online searching behavior and to monitor their performance by using a collaborative learning environment which tracks students' activities. This recipe is supported by a system that integrates a collaboration environment, a glossary tool, and an online tracking system, specifically created to meet the needs of teachers who teach translation and interpretation courses. Using this system, the teacher can take advantage of the dashboard visualizations to monitor students' activities and identify cases of low commitment or misunderstanding of the task so they can provide individual support to the students who need it.

Keywords LearnWeb · Learning platform · Log data analysis

7.1 Appetizer

In a modern learning scenario, informal learning activities like searching for materials on the Internet are increasing. If a teacher can track how students carry out their searches, the teacher can help them improve the quality of their search results. However, few tools are available that allow teachers to track their students' online search behavior efficiently. Such tracking would be particularly helpful in terminology work, as it involves searching the web for comparable texts and selecting those that provide useful translation equivalents.

Many studies have already used log data to analyze learning activities (Mazza and Dimitrova 2004; Mazza et al. 2012; Zhang and Almeroth 2010). Most of these studies use only the built-in logging facilities of tools like Moodle and WebCT, but many language-learning tasks require students to search for information on websites other than the tools used in the course, and these external actions cannot be logged by course management systems like Moodle. Other studies have used screen-

R. Jaakonmäki et al., *Learning Analytics Cookbook*, SpringerBriefs in Business Process Management, https://doi.org/10.1007/978-3-030-43377-2_7

capturing software like Camtasia and Abode Connect (Bortoluzzi and Marenzi 2017) to record students' learning activities, but since these recordings have to be analyzed manually, the number of subjects is limited by the number of evaluators. This approach can also be considered more obtrusive than server-side logging.

One way to address this problem is to create a tracking system that allows teachers to collect log data on the students' searching activity. This recipe is supported by a computerized system, LearnWeb, a learning platform that contains a glossary tool as well as a tracking system that are specifically created to support teachers who teach interpretation, and their students with regard to terminology. It takes advantage of data logs and learning dashboards to guide and support students' autonomous terminology work, inform the teacher of the students' activities and commitment to the task, and help the teacher to identify students in need of personalized remedial feedback. This recipe introduces the LearnWeb collaborative environment and its integrated tracking system in the Preparation section. The Serving section explains how the teacher can take advantage of the dashboard visualizations to monitor students' activities and identify cases of low commitment to or misunderstanding of the task so they can provide individual support to the students who need it.

Preparation time: 1 h

7.2   Ingredients

This recipe offers you a full collaborative learning environment that can be shared with your students, as well as integrated learning analytics tools with which you can monitor their activities. The basic ingredients are:

- *LearnWeb platform.*[1]

  - Described in Sect. 7.3.1, LearnWeb is freely available online and ready to use, including all functionalities described below.
  - Glossary (Sect. 7.3.2)
  - Proxy (Sect. 7.3.3)
  - Tracker (Sect. 7.3.4)
  - Learning analytics dashboard (Sect. 7.3.5)

- Access to the LearnWeb platform. Request a Registration Wizard to create an account personalized to your environment and customized for your class (Sect. 7.3.1).
- Your students' consent to use their data.

---

[1]http://www.learning-analytics-toolbox.org/tools/—LearnWeb

7.3   Preparation

The first step to cooking this recipe is setting up the learning environment on the LearnWeb platform. This section explains the basic functionalities that will allow you to prepare the environment and invite your students to start their activities in just a few minutes.

7.3.1   The LearnWeb Platform

LearnWeb (http://www.learning-analytics-toolbox.org/tools/) is a collaborative learning environment that allows users to share and work on user-generated resources or those collected from the web (Marenzi and Zerr 2012). It provides users with a search interface for resource discovery and sharing across Web 2.0 services like YouTube, Flickr, Ipernity, Vimeo, and Bing, as well as LearnWeb itself, so it can offer a personal Web 2.0 learning space.

The current release of the system provides several innovative features that are designed to support teachers and students in collaborative learning tasks:

- a personal learning space that offers a seamless overview of the entire set of learning resources distributed across various Web 2.0 repositories
- collaborative searching, sharing, and aggregation of learning resources
- annotation facilities like tagging and commenting that can help users discuss and select the best resources for their task and that support TED talk transcript annotation
- automatic logging of the students' actions

The current release also includes features that are designed and developed to support the learning activities of students who are attending a course on translation and interpreting. These features include a tool for the creation of glossaries, a tracking system for logging the students' searches on the web, and a learning analytics dashboard for improving the efficacy and effectiveness of the processes involved in the creation of personal glossaries. These three features are described in detail in the following sections.

But the first thing is to create a dedicated environment for your class by asking the LearnWeb team at learnweb-support@l3s.de to create a personalized Registration Wizard for you. When you receive the Wizard, you can use it to create your own user account and access a special instance of LearnWeb that you can customize for your students.

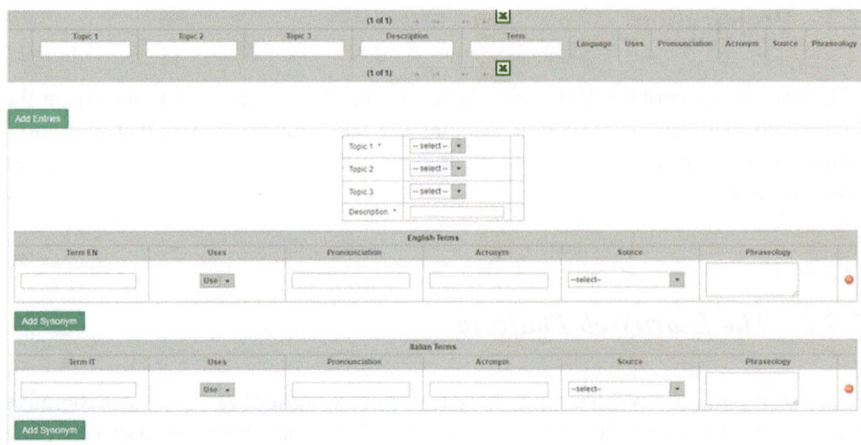

Fig. 7.1 Glossary interface

7.3.2   The Glossary

The glossary tool was designed to meet the needs of students who are studying to become interpreters. In LearnWeb, a glossary is a database that the system considers to be a type of resource that is on a par with images, videos, and text files. All of the data that users enter, along with their actions, such as viewing, adding, editing, and deleting entries, are logged and stored in a relational database with a time stamp.

A glossary can host an infinite number of entries, each of which is composed of at least two terms: the source term and its translation into the target language. Synonyms in the source and target languages may be added, so each glossary entry contains two or more terms. Each term in a glossary entry is accompanied by a series of attributes that correspond to given fields in the glossary interface; these attributes enrich the term with optional data. Currently, the glossary is set for the English-Italian language pair, but it can easily be adjusted for other language pairs.

LearnWeb glossaries can be filled in and consulted bidirectionally (e.g., from English into Italian and vice versa), can be personal and/or collaborative, and can be shared with other students/users. More details about the use of the glossary are in Sect. 7.5 (Serving learning analytics). Figure 7.1 shows the glossary interface for entering a new entry.

7.3.3   The Proxy

The main challenge in tracking users' Internet activities is that web servers' log files are usually accessible only to the server's owner. For example, the operator of a Moodle instance can easily log every request a user makes to his or her system, but when the user leaves the platform and opens an external web page to check a term at www.linguee.it, for example, all actions that occur on this page are outside the scope of the Moodle operator, so he or she cannot track what the user is doing without installing special tracking software on the user's computer.

To overcome this problem, we implemented a proxy service that keeps the user on our server, allowing us to log his or her activities. We make our system, the Web Analytics Proxy Service, available to other researchers under the domain www. waps.io. When a user follows a link to an external website like www.linguee.it, we redirect him or her to a subdomain on our server www.linguee.it.waps.io.

Our service will issue a request to the actual web resource at www.linguee.it and forward the response to the user. The same HTTP header information that we receive from the user (e.g., the preferred language, type of browser, cookies) is also sent to the actual resource. Otherwise, the user could retrieve a different response (e.g. another language) than he or she would get with the proxy service.

The returned page is likely to contain links to other pages on the same or other domains. To ensure that the user does not leave our proxy, which would interrupt the tracking, we modify all hyperlinks on the page to point to our proxy service.

7.3.4   The Tracker

The tracking framework allows teachers to track all pages a student views during a learning session without requiring changes to the students' computers. This system also tracks external websites like Wikipedia and Google. It is also not limited to classrooms, so students can access it from home and use online resources just as they normally do, so the system is much less obtrusive than previous approaches. We also track users' mouse movements and keyboard inputs to detect when they actively interact with a web page.

The proxy server can create logs that contain simple information, such as the browser model, the URL, and the date when the user visited a website. To gather more fine-grained information, we use JavaScript to record all inputs the users make, including mouse movements, scrolling, clicking, and typing. We also record when the browser window loses the focus, that is, when the user switches to another browser tab or puts the whole browser window into the background. For each input action, we record the time and the curser position on the page.

To limit the log size and the amount of transferred data, we record up to three cursor positions per second while the mouse is moving. The log data is accumulated into batches and sent asynchronously to our tracking server so the tracker does not influence the browsing experience. This data is used to calculate how active a user was on a page, so we treat all subsequent actions that take place within a sliding window of five seconds as one continuous action and assume that the user was active during the whole time between the first and last actions. For each log entry, the system can show statistics, including how much time the user spent on a site and how long he or she was active (moved the mouse, scrolled, clicked, or typed).

For this purpose, the system shows the page at exactly the same window size at which the user viewed it and can visualize the recorded mouse cursor movements. At the top of the window is a timeline that shows when the user was active during his or her session on that page: Yellow indicates that the user moved the mouse, and red shows that he or she scrolled or clicked. This timeline is shown only to developers and researchers, not to students.

7.3.5   The Learning Analytics Dashboard

The dashboard is a graphic and interactive representation of the students' data based on a selection of log data from the glossary activities. Data can be selected to monitor:

1. **General activity on the glossary task**: The dashboard shows general information about the students' activities, the number of glossary items entered, the number of terms, and the number and types of sources.
2. **Ability to organize the glossary**: The dashboard provides information about whether, to what extent, and how the student completed the compulsory and optional fields, including summary and percentage data about the number and types of fields filled in, a list of the descriptions entered, their length, and the language used in the description field.
3. **Ability to search the web for information**: The dashboard shows the students' preferred sources and a list of all the web addresses that were opened and tracked by the proxy.

More details about the dashboard visualization can be found in Sect. 7.5.

7.4   Cooking

If you are a teacher and you want to cook this recipe, these are the steps you need to follow:

- Step 1

  - Access the LearnWeb environment created for your classroom.
  - Ask for a customized Registration Wizard for your course (Sect. 7.3.1).
  - Design the environment according to the learning activities (e.g., folders, group of resources).
  - Share the Registration Wizard with your students and ask them to create an account.

- Step 2

  - Assign tasks to your students based on the learning goals.

- Step 3

  - Let students work in LearnWeb.
  - From the very beginning, inform the students that the system will track all their actions and all the websites they visit. We suggest asking explicit permission for that.

- Step 4

  - Look at the dashboard visualizations to monitor students' activities so you can give them personal feedback and improve the course design.

7.5 Serving

Let's work with a real example.

- (See Step 1) Imagine you are a teacher of a translation and interpreting course at a university and you want to teach an MA module that trains students in consecutive interpreting without notes for the English-Italian language pair. The module focuses on medical language. You can ask the LearnWeb team (at learnweb-support@l3s.de) for a Registration Wizard so you can create accounts to access an environment that is customized for your course.
- (See Step 2) Students create their personal accounts using the Registration Wizard and access the LearnWeb environment you prepared for the course. As a learning activity, students will create personal medical glossaries using the glossary tool in LearnWeb. You design the learning activities starting from a TED talk on medicine and from other materials previously used in class so the students have with a concrete starting point from which to identify candidates for their glossaries.
- (See Step 3) After entering the LearnWeb system, students can start annotating the assigned TED talk and filling their first few terms into the personal glossary (Fig. 7.2). They are allowed to search in information sources that get the required information, such as:

  Description: A short definition or description of the concept that is linguistically realized by the given terms. This field plays multiple roles in the glossary rationale: It is a way to draw the student's attention to the meaning of the terms, rather than the terms themselves; it can be used to retrieve terms even when one does not remember them, as this field is a searchable text box; and from a technical point of view, the description field keeps the source and target terms, their synonyms, and all other attributes together to form a single entry.
  Term EN: Where the student enters the English term.
  Term IT: Where the student enters the Italian term.

Topic 1 and **Topic 2**: These fields help the student classify the entries according to a logical and functional ontology. **Topic 3** is an open text box. These fields and menus can be customized to the teacher's needs upon request.

Sources menu: The menu invites the student to specify the primary source of information he or she used to fill in the glossary fields. The menu lists the following options: Wikipedia, encyclopedia, monolingual dictionary, bilingual dictionary, glossary, scientific/academic publication, institutional website, Linguee or Reverso, patients' websites and blogs, and other.

The sources menu may seem redundant since an automatic tracking feature logs all of the students' web searches. However, it is important for students to think about the type of source(s) they use.

The entire glossary, as well as the filtered results, can also be downloaded in Excel format by clicking on the Excel icon and printing it out if necessary. Figure 7.2 shows a few entries taken from a student's glossary.

After a couple of hours during which each student creates a personal glossary and works on it in class, the students can be invited to continue building their glossaries from home as part of their individual preparation for the final exam.

- (See Step 4) At the end of the activity, the teacher can use the dashboard to visualize a quantitative analysis of the students' online behavior in terms of the resources they accessed. Dashboard data can help the teacher identify cases of low commitment or misunderstanding of the glossary task so he or she can provide individual support to the students who need it.

The following paragraphs discuss how the dashboard looks, going from more general to more detailed views of the students' actions. The data is taken from a real teaching scenario carried out at the University of Salento, Lecce, in spring 2017.

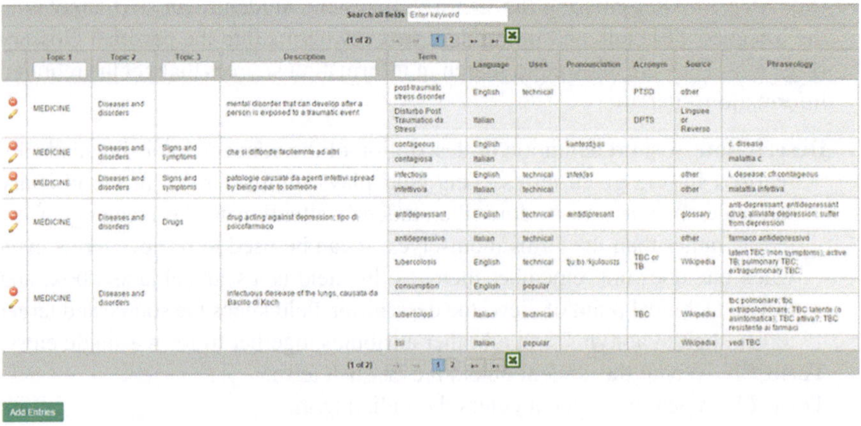

Topic 1	Topic 2	Topic 3	Description	Term	Language	Uses	Pronunciation	Acronym	Source	Phraseology
MEDICINE	Diseases and disorders		mental disorder that can develop after a person is exposed to a traumatic event	post-traumatic stress disorder	English	technical		PTSD	other	
				Disturbo Post Traumatico da Stress	Italian			DPTS	Linguee or Reverso	
MEDICINE	Diseases and disorders	Signs and symptoms	che si diffonde facilemnte ad altri	contageous	English		kən'teɪdʒəs			c. disease
				contagiosa	Italian					malattia c
MEDICINE	Diseases and disorders	Signs and symptoms	patologie causate da agenti infettivi spread by being near to someone	infectious	English	technical	ɪn'fekʃəs		other	i. desease; cfr.contageous
				infettivo/a	Italian	technical			other	malattia infettiva
MEDICINE	Diseases and disorders	Drugs	drug acting against depression; tipo di psicofarmaco	antidepressant	English	technical	æntɪdɪpresənt		glossary	anti-depressant, antidepressant drug, alliviate depression, suffer from depression
				antidepressivo	Italian	technical			other	farmaco antidepressivo
MEDICINE	Diseases and disorders		infectuous desease of the lungs, causata da Bacillo di Koch	tubercolosis	English	technical	tju:bɜ:ˈkjuləʊsɪs	TBC or TB	Wikipedia	latent TBC (non symptoms); active TB; pulmonary TBC; extrapulmonary TBC;
				consumption	English	popular				
				tubercolosi	Italian	technical		TBC	Wikipedia	tbc polmonare; tbc extrapolomonare; TBC latente (o asintomatica); TBC attiva?; TBC resistente ai farmaci
				tisi	Italian	popular			Wikipedia	vedi TBC

Fig. 7.2 Example of a student's glossary

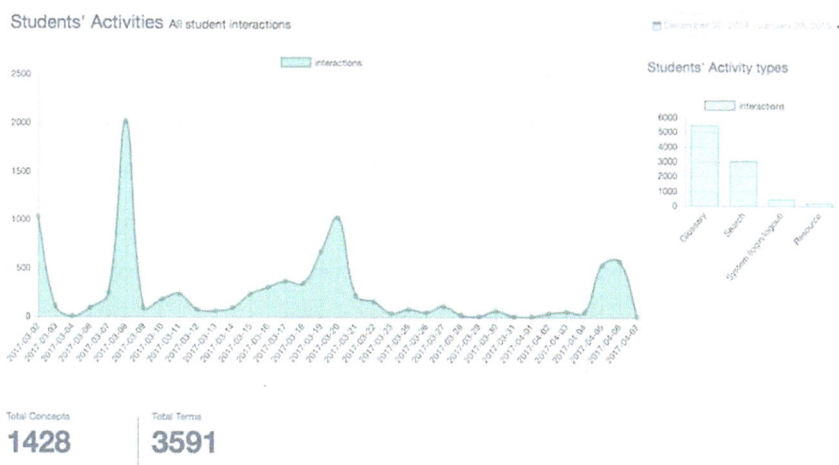

Fig. 7.3 Teacher's dashboard—Class view of the students' activities

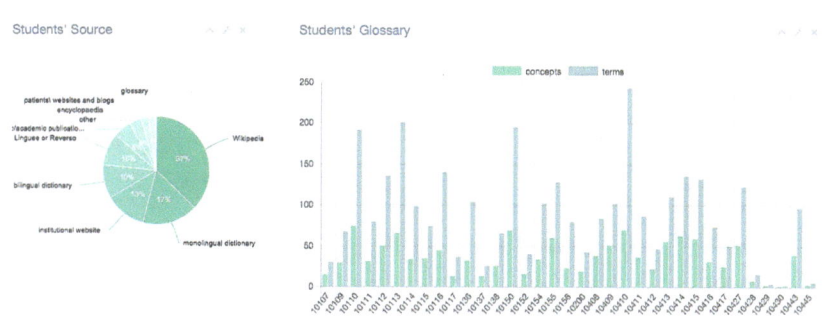

Fig. 7.4 Teacher's dashboard—Class view of the students' sources of information

The teacher's dashboard opens on a general view of the group, with particular attention to the activities undertaken in building the glossary (Fig. 7.3) and their primary sources of information (Fig. 7.4).

The general dashboard view of the students' activities (Fig. 7.3) shows that most of the students were not expanding their glossaries daily, as the teacher suggested, but only when spurred to do it. The glossary interface was explained in class on March 2, an entire class lesson was dedicated to glossary-building on March 8, and the deadline for a glossary homework assignment was March 21.

However, there are exceptions, such as student 10,113, who worked on the glossary on 21 days out of 39 (Fig. 7.5).

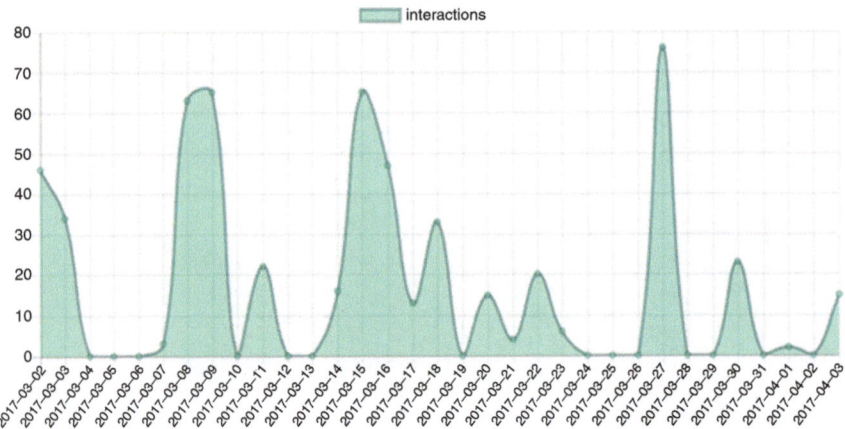

Fig. 7.5 Exceptional behavior—Activity view of student 10,113

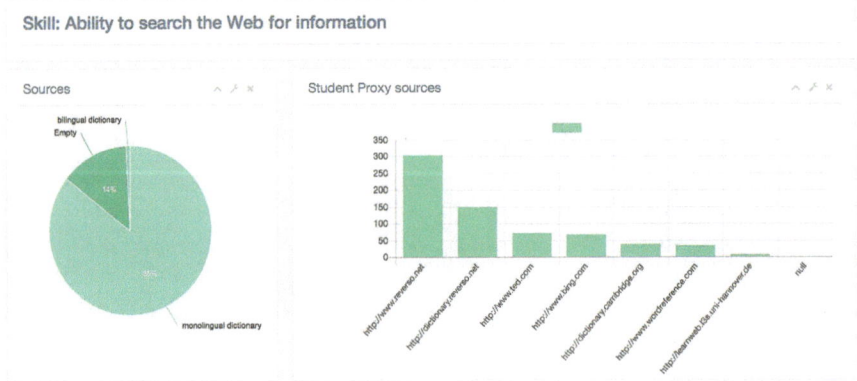

Fig. 7.6 Teacher's dashboard—View of a student's ability to search the web for information

The teacher can zoom in on individual students by clicking on the student ID to access views of the student's activities, similar to those in Fig. 7.5, and views of the student's ability to organize the glossary and search the web for information (Fig. 7.6).

A general view of the number of terms the students entered and the types of fields they filled in (Fig. 7.7) provides an overview of the students' commitment to the glossary task.

Figure 7.7 shows that student 10,113, for example, entered a high number of terms (228) and also completed 63.42% of the optional fields. Student 10,430, while he or she entered few terms, completed each of them with 100% of the possible details. (The absence of acronyms in this case, which would lower the completion average to 80, is considered irrelevant since not all medical terms have corresponding acronyms.) Both types of behavior suggest high commitment to the

Filled in fields ︿

	terms	pronunciation	acronym	phraseology	uses	source	AVG %
10110	211	27.49%	9.95%	30.81%	86.73%	59.24%	42.84%
10112	162	46.91%	4.32%	70.99%	96.91%	91.98%	62.22%
10113	228	35.09%	18.86%	69.30%	96.49%	97.37%	63.42%
10150	214	17.76%	12.15%	36.92%	86.45%	56.07%	41.87%
10410	243	4.12%	3.70%	1.23%	22.22%	41.15%	14.49%
10430	34	100.00%	0.00%	100.00%	100.00%	100.00%	80.00%

Fig. 7.7 Number and types of fields filled in by the students

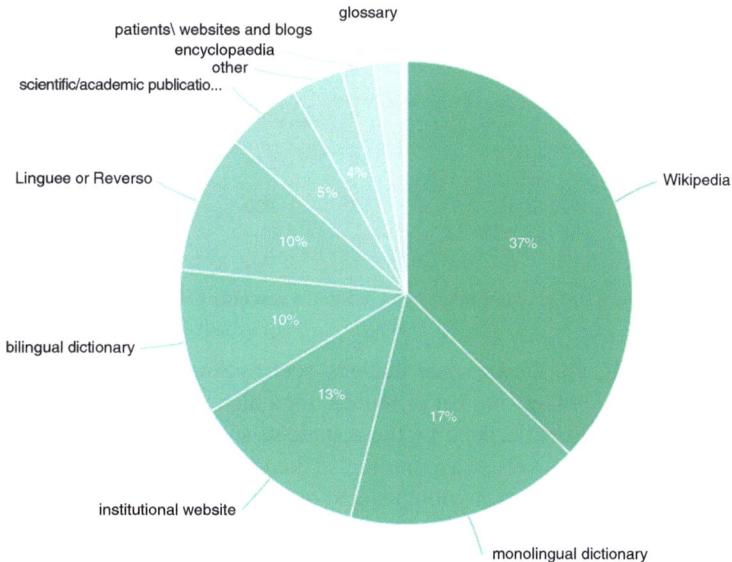

Fig. 7.8 Group view of the data students entered in the resource field

task. By contrast, student 10,410 entered many terms (243) but did not complete the optional fields in the glossary (overall percentage was 14.49%), which suggests a low level of commitment and possibly also little understanding of the importance of contextual information.

Let us now zoom in on specific glossary fields. The dashboard in Fig. 7.8 shows that, in the resource field, students primarily consulted content-based resources (62.9%) like Wikipedia and other types of encyclopedias, institutional websites, scientific publications, and in some cases language-based resources (37.1%) like dictionaries and parallel corpora (e.g., Linguee and Reverso). Among the content-based resources, Wikipedia was by far the preferred one resource. This type of profile shows that the group understood the purpose of the glossary and the importance of extended reading about a topic.

Summary			^
	sources	**terms**	**entries**
10413	1	110	55
10430	1	2	1
10429	1	4	2
10443	3	97	39
10411	3	87	36
10152	3	40	16
10111	3	32	14
10117	3	36	13
10428	3	16	8
10113	7	200	65

Fig. 7.9 General view of the data students entered in the resource field

However, as Fig. 7.9 shows, some students, such as Students 10,413, 10,443, and 10,111, used a very limited number of sources for a high number of entries. A zoom in on the type of sources declared by these students shows that student 10,413, for example, resorted exclusively to a monolingual dictionary. Similarly, student 10,111 made ample use of dictionaries (bilingual and monolingual) and very limited use of content-based sources and a high percentage of entries for which no source was specified (Fig. 7.10).

Such visualizations allow the teacher to analyze students' activities, search behavior in detail, and organize individual conversations with some students where needed. For example, student 10,413 and student 10,111 confirmed during such conversations that they had not understood that a glossary is a way to learn about a topic and, more generally, that they had not understood the importance of content and context in interpreting.

Other students, such as student 10,113, entered several types of sources, although with a clear preference for Wikipedia (Fig. 7.11).

Curious to see which websites student 10,113 had visited and to what extent they coincided with what the student had declared, the teacher checked the student proxy sources list (Table 7.1) and manually classified them by type. This classification showed that the student had spent a lot of time searching for information in websites of type "other" (31% of all visited sites; these included newspapers, sites for the medical profession, promotional sites of pharmaceutical products or natural

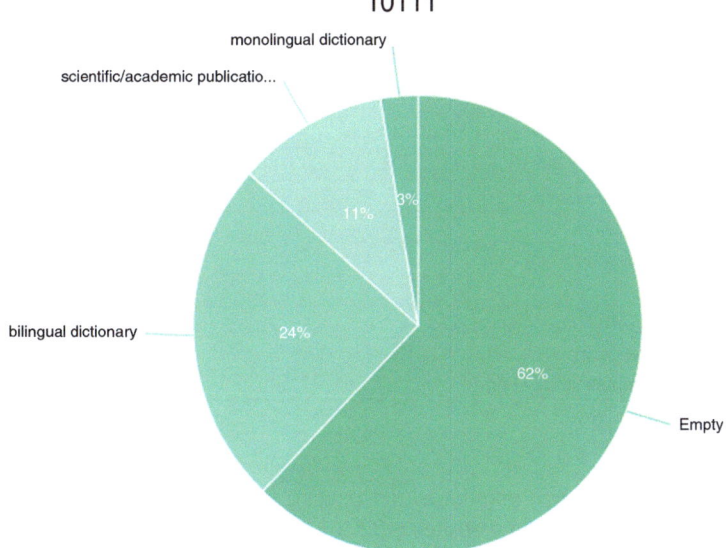

Fig. 7.10 Data that student 10,111 entered in the resource field

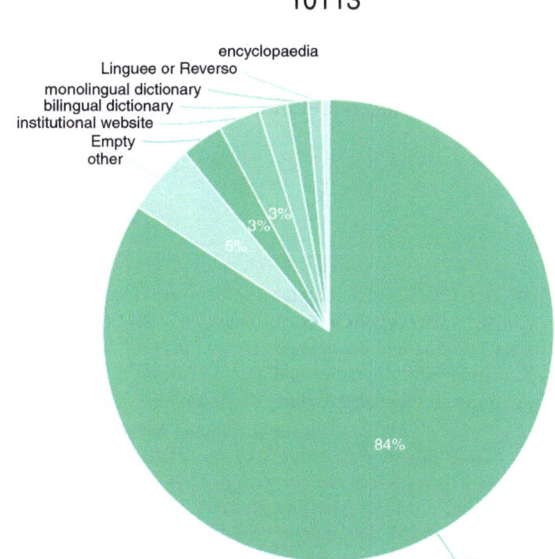

Fig. 7.11 Data that student 10,113 entered in the resource field

Table 7.1 Student 10,113—Selected lines from the student proxy sources list in the dashboard

Website	Occurrences
http://it.wikipedia.org	1405
http://www.britannica.com	794
http://www.merriam-webster.com	680
http://www.medicinenet.com	574
http://www.wordreference.com	451
http://www.youtube.com	450
http://www.differencebetween.com	383
http://prezi.com	328
http://www.organsofthebody.com	322
http://www.diabetes.org	311
http://www.medicalnewstoday.com	215
http://www.nal.usda.gov	156

supplements, and science popularization pages), followed by monolingual dictionaries (17.6%), institutional websites (13.5%), Wikipedia (19.7%), and other encyclopedias (11.6%).

The search strategies of student 10,113 are in keeping with the teacher's suggestions to search in a wide range of websites and to focus on content-based sites rather than language-based ones. However, the data entered in the sources field were the student's declaration that, all in all, Wikipedia was the most profitable source, probably because it offers basic scientific information in a highly structured way. The teacher can use this type of information, for example, to open a class discussion about the range of available sources and their advantages and disadvantages.

References

Bortoluzzi, M., & Marenzi, I. (2017). Web searches for learning: How language teachers search for online resources. *Lingue e Linguaggi Lingue Linguaggi 23*, 21–36. Retrieved from http://siba-ese.unisalento.it/index.php/linguelinguaggi/article/viewFile/17056/15780

Marenzi, I., & Zerr, S. (2012). Multiliteracies and active learning in CLIL—The development of LearnWeb2. 0. *IEEE Transactions on Learning Technologies, 5*(4), 336–348.

Mazza, R., & Dimitrova, V. (2004, May). Visualising student tracking data to support instructors in web-based distance education. In *Proceedings of the 13th International World Wide Web conference on Alternate Track Papers & Posters* (pp. 154–161). ACM.

Mazza, R., Bettoni, M., & Far'e, M., & Mazzola, L. (2012). Moclog—monitoring online courses with log data. In S. Retalis & M. Dougiamas (Eds.), *1st Moodle Research Conference Proceedings* (pp. 132–139). Retrieved from https://research.moodle.net/54/1/17%20-%20Mazza%20-%20MOCLog%20-%20Monitoring%20Online%20Courses%20with%20log%20data.pdf

Zhang, H., & Almeroth, K. (2010). Moodog: Tracking student activity in online course management systems. *Journal of Interactive Learning Research, 21*(3), 407–429.

Glossary

Analytics Describes the subject "data analytics," where data is analyzed to identify patterns that can help a user make better decisions.

Analytics technologies Technologies like data visualization and predictive analytics that are used in analytics to support decision-making.

API The abbreviation for Application Programming Interface, which provides guidelines for developers on how to make an application for a certain environment.

Backup Securing data by creating duplicates stored at different places.

Big data A huge amount of structured or unstructured data, often gathered from a wide variety of sources.

Big data analytics An analysis of big data to generate business value from it.

Central data store The place where data is stored; it can be stored in any format and often can be found in a database or similar kind of storage.

Classification Categorizing objects for simplified management of information.

Computerized system A system in which all necessary software and hardware is already in place and well-documented.

Csv file An abbreviation for Comma-separated values file, which is a file that is organized in individual columns separated by a comma and rows separated by lines.

Data abuse Misuse of data.

Data client The party who receives the data from data subjects.

Data collection A structured assemblage of data to gain insight into a certain topic.

Data log A set of historical data activity related to who used the software and how it was used collected by a computer (e.g., through sensors) and often automatically recorded and stored by educational application software, such as Moodle.

Data management Handling of available data, which includes processes like creating, storing, editing, and deleting data.

Data minimization The effort to protect individual data by storing only the data that is important, which, according to the EU GDPR, is the case if data is adequate, relevant and limited to certain purposes.

© The Author(s) 2020   89
R. Jaakonmäki et al., *Learning Analytics Cookbook*, SpringerBriefs in Business
Process Management, https://doi.org/10.1007/978-3-030-43377-2

Data owner The entity that is in charge of handling access to data and the data's status.

Data privacy Also called information privacy, refers to protection of users' or organizations' individual data, including the boundaries of what data can be shared with third parties. Also called information privacy.

Data protection The means used to keep data safe from manipulation and unauthorized access.

Data science A field that deals with handling data and applying analytics methods to identify its potential to create value.

Data subjects A term from the EU GDPR that refers to the person whose data is being considered for use.

Data-driven innovation Innovation based on available data that is used for further or new development of a product.

Dataset A collection of data, examples of which include students' course performance-related data.

Digital rights Regulations concerning individual rights to access and use digital devices or digital media, including the rights to create, upload, or download digital media.

Digital technologies Devices that require the use of binary code to function, including all computer-based products.

Digital trace What is left behind from users when they use mobile and digital devices; a reconstruction of the actions performed with the device, collected by the application, and stored in a database.

Drop-out alert system A system that enables a teacher to react before a student faces the risk of failing or dropping out.

Educational data science Applying data science in the educational domain.

Electronic communication Communicating with only electronic devices without direct contact or geographical limitations.

Inquiry-based learning A particular form of learning in which the learner actively asks questions.

Key performance indicators Measures that are used to evaluate the success of performance.

Landing page The page to which one is redirected to when following a link.

Learning analytics A stream in research that includes collecting, analyzing, and presenting data from learners or learning situations to improve teaching and learning.

Log A record of activity, names, etc.

Log data See data log.

Machine learning Method in which data is given to artificial intelligence (a computer system) that learns patterns from the data.

Macro-level Used to describe something on an abstract, undetailed level.

Meso-level Used to describe something on an average level.

Metacognitive awareness Ways to help students to increase their learning performance.

Meta-data Data that describes other data (e.g., name, date, location).

Micro-level Used to describe something on a highly detailed level.

Multifaceted data sets Data sets with multiple features.

Multi-faceted tracking Tracking data from many perspectives.

Paradigm shift Severe change in the usual approach to something.

Personal data Information about an individual/natural person.

Privacy concerns Fear of abuse of data by a third party.

Privacy protection Protection of personal data.

Proxy service A system that enables a user device to connect to the internet with the help of a proxy server.

Regulatory Based on rules or legislation.

Relational database Structured storage of data in tables of rows and columns.

Search interface User view of search window from a database.

Self-regulatory regimes A regime that sets its own standards and rules.

Sensitive data Information that should be protected from unrestricted use.

Server A computer program that offers services (e.g., data, functionality) to other computer programs

Server's owner The entity that is responsible for a server.

Social network analysis Examination of relationships on social networks.

Statistical analysis techniques Methods used for statistical evaluation.

Sub-group A group that is part of a larger, superordinate group.

Superordinate groups A group that is higher on a hierarchical level than other groups (often has sub-groups).

Technical innovation A new technical development

Temporality of data Refers to the timeframe within which data is bounded or collected.

Tracking Pursuing the (digital) tracks left by a user.

Transparency Clear visibility of official actions to external entities.

URI An abbreviation for Uniform Resource Identifier, which allows certain resources to be tracked and identified using a unique code.

Validity The degree to which information is accurate and believable.

Venture capital Invested money or expertise into a project that involves a certain risk.

Visual analytics A field of analytics that seeks to visualize the results of data analytics in an interactive way.

Widget A tool in the user interface that is used for a specific purpose and is often based on an application.

xAPI An abbreviation that stands for Experience API; a specification that allows learning systems to speak with each other and understand the data in the same way.

Zip file A file reduced in size for easier transfer and handling of data.